A HUNTINGTON LIBRARY CLASSIC

Huntington Library Classics

Autobiography of a Los Angeles Newspaperman, 1874–1900 by William Andrew Spalding

The Butterfield Overland Mail: Only Through Passenger on the First Westbound Stage by Waterman L. Ormsby

The Cattle on a Thousand Hills: Southern California, 1850–1880 by Robert Glass Cleland

Charles F. Lummis: Editor of the Southwest by Edwin R. Bingham

The Cosmographical Glass: Renaissance Diagrams of the Universe by S. K. Heninger Jr.

The Health Seekers of Southern California, 1870–1900 by John E. Baur; introduction to the second printing by Robert G. Frank Jr.

Ho for California! Women's Overland Diaries from the Huntington Library edited by Sandra L. Myres

The Irvine Ranch by Robert Glass Cleland

Jack London and the Klondike: The Genesis of an American Writer by Franklin Walker

Juan Rodriguez Cabrillo by Harry Kelsey

Mexican Gold Trail: The Journal of a Forty-Niner by George W. B. Evans

A Mormon Chronicle: The Diaries of John D. Lee, 1848–1876

Sir Richard Burton's Travels in Arabia and Africa: Four Lectures from a Huntington Library Manuscript by Sir Richard Burton

Thirty Explosive Years in Los Angeles County by John Anson Ford; introduction to the second printing by Michael R. Adamson

A Victorian Gentlewoman in the Far West: The Reminiscences of Mary Hallock Foote

THE HEALTH SEEKERS

OF SOUTHERN CALIFORNIA, 1870–1900

by John E. Baur

INTRODUCTION TO THE SECOND EDITION

by Robert G. Frank Jr.

THE HUNTINGTON LIBRARY, SAN MARINO, CALIFORNIA

Henry E. Huntington Library and Art Gallery
1151 Oxford Road, San Marino, California 91108
www.huntington.org

First paperback printing 2010
Printed in the United States of America

Cover design by Doug Davis
Images from items in the Huntington's collections are reproduced by permission.

Library of Congress Cataloging-in-Publication Data
Baur, John E.
The health seekers of Southern California, 1870/1900 / by John E. Baur.
p. ; cm.
Includes bibliographical references and index.
ISBN 978-0-87328-225-3 (pbk. : alk. paper)
1. Health resorts—California, Southern—History—19th century. I. Title.
[DNLM: 1. Health Resorts—history—California. WB 760 AC2 B3h 1959a]
RA807.C2B36 2010
613'.122097949—dc22

2008027840

Contents

INTRODUCTION TO THE SECOND PRINTING

Tuberculosis and Medicine in the Era of Health Seekers

by Robert G. Frank Jr.

When the twenty-four-year-old UCLA graduate student John Baur began, about 1947, to collect information on the "health rush" to nineteenth-century California, pulmonary tuberculosis was embedded in the collective consciousness of Southern California. One reminder was institutional. Scattered across the region were numerous sanatoriums, large and small, that provided residential care to seriously ill patients. The Barlow Sanatorium, in Los Angeles's Chavez Ravine, was founded in 1902 by physician Walter Jarvis Barlow (himself ill with tuberculosis) for the intensive care of indigent "lungers" in the city. Francis Marion Pottenger, whose wife had the disease, established the Pottenger Sanatorium in the foothills of nearby Monrovia in 1903 for private patients; Baur interviewed him in 1951. The Jewish Consumptive Relief Association founded the Los Angeles Sanatorium in 1913 in the adjoining town of Duarte, likewise taking advantage of the slightly higher elevation and drier climate. Forging connections with Hollywood's nascent movie industry, the Los Angeles Sanatorium became nonsectarian, and grew enormously in the 1920s and 1930s as a Warner Bros. favorite charity. In 1946, the year before Baur began his investigation, it started to shift its focus gradually to cancer and soon emerged as the "City of Hope."[1] Olive View Sanatorium, in the northwest end of the San Fernando Valley, was built in 1920 by Los Angeles County for the residential treatment of pulmonary tuberculosis. Its pathology laboratory, directed by Dr. Emil Bogen, was known internationally for its studies of the tubercle bacillus. Birmingham Veterans Administration Hospital, in Van Nuys, had large wards for extremely disabled victims, particularly paraplegics and tuberculars. Overall, tuberculosis patients occupied 2,773 of the 10,082 hospital beds in the area.[2] Binding this network together was the very active and influential Los Angeles County Tuberculosis and Health Association.

At the other end of the spectrum was popular medicine, such as that purveyed by the daily medical column "Here's to Health" in the *Los Angeles Times*. Hardly a week went by in 1947 that William Brady, M.D., didn't dispense education and advice about tuberculosis. Should a reader marry his fiancée, who is infected? Check with your doctor: if the disease is still active, marriage would be a bad gamble; if it has been arrested, then "marry, multiply and be happy."[3] Has your child's skin test in school proved positive for tuberculosis? Then a chest X-ray and detailed examination are in order.[4] Brady explained patiently that while many were infected with the tubercle bacillus, active cases occurred only in the few whose immunity could not resist it.[5] No, he said, you cannot get tuberculosis from objects handled by patients, or from casual visits with them.[6]

Organizations official and unofficial worked ceaselessly to prevent and detect the disease. From Sacramento the newly re-elected governor, Earl Warren, used his inaugural address to urge, among other improvements, greater state aid to the counties in their fight against tuberculosis.[7] The California Medical Association, while exhorting its members to repel the specter of socialized medicine, recognized that the profession had to cooperate with the government in finding solutions to "the triple threat problems of tuberculosis, cancer, and heart disease."[8] In Los Angeles County, the Tuberculosis and Health Association put a portable X-ray unit into service; the *Times* naturally gave full news coverage when the machine was used for free screening at the newspaper's Boys Club.[9] Soon it was put to mass use in making chest films of more than two thousand students at UCLA (with John Baur perhaps among them).[10] The unit was similarly used to screen teachers and school employees for the disease—a practice very much in line with the aim of the Los Angeles Parent-Teacher Association to have tuberculosis testing in every school.[11] Even non-students could not miss the public-service newspaper announcements exhorting in bold type, "Fight Tuberculosis. An X-ray today can bar TB tomorrow!"[12]

Costs of both hospitalization and public education were underwritten through benefits and volunteer work. Film and radio stars such as Eddie Cantor, Jimmy Durante, Danny Kaye, and Dinah Shore lent their dazzle to the Cocoanut Grove nightclub when the Los Angeles Sanatorium held its gala fundraiser there in March 1947.[13] At Legion Stadium, the Sportsmen's Club sponsored an annual boxing match called, appropriately, "The Fight for Life," and it yielded more than $100,000 the previous year to combat tuberculosis.[14] Rather less grand were the teas organized by Hollywood socialites to thank volunteers of the Tuberculosis and Health Association, or the luncheons sponsored by the Republican Women's Club to collect gifts for children of the tuberculous patients in the Veterans' Administration wards at Birmingham Hospital.[15] Most of all, the enterprise of tuberculosis awareness was underwritten by the annual sale of Christmas Seals for the National Tuberculosis Association, a tradition since 1907.[16] Seals were sold for a penny apiece on the streets, door to door, and—most picturesquely—from a booth at the corner of Hollywood and Vine, where a choral group sang carols every evening.[17] Newspaper stories that ran between Thanksgiving and Christmas tracked the growing sum raised locally by volunteers of the Los Angeles Tuberculosis and Health Association in support of education, patient care, and research.[18]

These efforts gained vigor through the increasing public awareness that a cure, a "miracle drug," had been discovered. Streptomycin, manufactured by Merck & Co., was in limited supply, and until only recently had been controlled by the U.S. Army. Civilians first heard of it as an antibiotic effective against urinary tract infections, typhoid fever, and rarer diseases such as bubonic plague and Malta fever (brucellosis). Only in mid-1946 did Southern Californians first read of the drug's use to cure tuberculosis of the bladder.[19] Thereafter it was used to treat tubercular meningitis, and finally was released to the public in the autumn of 1946.[20] In November of that year, the story of five-month-old Laura Lee Laird captured the attention of the entire city, when the drug saved her from a virulent systemic tubercular infection.[21] A few months later the drug saved another Los Angeles child, fifteen-month-old Harrel Ray Clark Jr., from the same lethal systemic (miliary) tuberculosis.[22] Thereafter Emil Bogen of Olive View Sanatorium became a spokesman for the use of streptomycin for pulmonary tuberculosis.[23] He received one of only three $25,000 grants from the fledgling National Institutes of Health to study its effects.[24] The drug was used in wider and wider circles, and became for almost a decade the mainstay of tuberculosis treatment.

Large sanatoriums, popular fears and medical advice, screening and testing programs, volunteers and fundraising campaigns, a "miracle drug"—all bespeak a collective consciousness and a collective memory that have now all but disappeared. Yet to read John Baur's elegant and beguiling book without recalling that public preoccupation with tuberculosis would be to lose much. Here I wish not to extend or rewrite his vivid narrative of Southern California culture and events from 1870 to 1900, but rather to set that era against another, and broader, narrative. As Americans in the 1940s well knew, those last three decades of the nineteenth century had seen a complete revolution in the medical understanding of pulmonary tuberculosis. Physicians practicing around 1860 believed that the disease—much more often dubbed "phthisis" or "consumption"—was constitutional and frequently hereditary, that it consisted of lesions called "tubercles" that ravaged the lungs and other organs, and that it was to be treated in the home by the normal mainstays of medical therapy. By 1900 pulmonary tuberculosis was universally recognized as an infectious disease caused by a bacterium that could be seen and isolated, and that caused the formation of the characteristic tubercles. Its tracks could be seen antemortem via roentgen-ray films, the presence of the disease could be detected using tuberculin in a skin-reactivity test, and actively infectious cases could be identified by the presence of tubercle bacilli in the sputum. The most effective treatment was to remove the actively infectious individuals to a sanatorium, giving them rest, sunshine, open air, and a regulated diet until the disease was arrested; the patients could then safely return to society. Understanding this revolution in medicine is fundamental to appreciating the era of health seeking. This historical narrative about the treatment of the disease requires, first, a brief account of its medical aspects.

Twentieth-Century Facts and Nineteenth-Century Reflections

The story begins—but certainly does not end—with the tubercle bacillus, *Mycobacterium tuberculosis*.[25] It is relatively small as bacterial pathogens go, about 0.5 μm (micrometers) by 3 μm, considerably smaller than a round red blood cell of about 6–8 μm across. Its small size and translucent appearance make it difficult to see unstained under the microscope, especially when it is among other cells. *M. tuberculosis* is derived from a group of bacteria that inhabit the soil, but the tubercle bacillus and its kin have specialized over tens of thousands of years to live in warm-blooded hosts, especially humans. It is unusually slow growing, dividing only every fifteen to twenty hours, in comparison to every twenty minutes for many other bacterial parasites. It is also extremely robust, withstanding weak disinfectants, living for weeks in a dried state, and—most important of all—resisting digestion when it is engulfed by the specialized white blood cells (macrophages) that routinely devour ordinary invasive bacteria. Tubercle bacilli can even divide inside their predators and destroy them. *M. tuberculosis* is a perfect pathogen for a chronic disease.

The tubercle bacillus spreads through droplets. When sufferers with active pulmonary tuberculosis cough, sneeze, or even talk, they aerosolize thousands of bacteria from the lungs into minute droplets of moisture, each with a few bacilli. These particles are so small that they dry rapidly, and they can hang in the air for hours. Transmission of the disease is determined by how dense the cloud is, how frequently it occurs, and how long a susceptible person is in contact with it. When droplet nuclei are inhaled, almost all are trapped by laryngeal mucus or carried away by the rhythmic sweeping of the cilia that line the bronchi. A few, however, can penetrate deeply into the lungs, and it only takes a few to start the infection.

Tuberculosis is thus very much a disease that tends to afflict humans living closely together in poorly ventilated spaces. The large industrial and commercial cities of nineteenth-century America—Boston, New York, Philadelphia, Cincinnati, Chicago, St. Louis—were perfect seedbeds. The factory proletariat and day laborers described and photographed by Jacob Riis[26]—with entire families squeezed into a single room because lodgings were so expensive—could not help but infect one another. Even the comfortable middle classes of Europe and America, who could afford larger apartments, often lived in close daily contact in parlors and kitchens whose windows were securely fastened, both to conserve heat and to exclude "unhealthy drafts." In sparsely populated rural areas, families often lived, ate, and sometimes slept together in a few rooms, especially during long winters. Given the continual presence of active cases, virtually everyone in nineteenth-century America was probably exposed to, and infected with, the pathogen.

Once the bacilli reach the terminal air passages deep in the lungs, they lodge there and begin a slow process of multiplication. But they are by no means unopposed. Macrophages ingest the invasive bacilli, and other parts of the immune system react to neutralize the infection. Antibodies mark the onset of this process. In a healthy person with normal lungs, the immune system contains the infection in the vast majority of cases. The bacilli are walled off, and the result is a small, translucent particle that is the tubercle. The affected person is seldom even aware of the triumph and perceives only a malaise and transient fever. Not all of the bacilli, however, are destroyed. They linger in a state of dormancy, ready to be reactivated when the immune system weakens (as with old age or sickness), or when freshly stimulated by the ingress of other bacteria. In less than 5 percent of cases, immune reactions are not successful in containing the bacilli, and the person proceeds immediately to an active case of pulmonary tuberculosis. If left untreated by antibiotics, it will often lead to death within two to five years—or in the case of young children, frequently within a few months.

Whether someone successfully fights the disease to a draw, or succumbs to it, depends on the person and their lungs. Genetic endowment may play a small part. Nutrition is much more important. An imbalanced diet, especially one too high in carbohydrates, and with little protein from meat, fish, or dairy products, more than doubles the chance of active disease. Alcohol abuse contributes almost as much risk. Strenuous work, especially combined with a poor diet, increases the liability for an active case. Age is also crucial, with the highest incidence of diagnosed tuberculosis found in late adolescence and early adulthood. For reasons that are unclear, young women are more prone to tuberculosis than young men. Most deleterious of all pre-existing conditions are lung inflammations, whether from concomitant diseases like bronchitis and pneumonia, or from long-term irritants like inhaled particles. Tuberculosis is a disease of many causes; the presence of *M. tuberculosis* is necessary, but virtually never sufficient.

Even beyond crowding, nineteenth-century America served up causes in abundance. Men routinely worked fifty to sixty hours per week in physically strenuous occupations. The diet of the poor tended toward carbohydrates, such as bread and potatoes, and beer. Whisky was cheap and ubiquitous. Even the fare of the better off, while probably containing adequate protein, was washed down with copious amounts of alcohol. Neither group ate many fruits or vegetables.

America was a young society, with many exactly at the age most prone to active pulmonary tuberculosis. More important were the multiple assaults on Americans' lungs. Miners had silicosis, which raised the chance of contracting consumption thirtyfold. Workers in textile factories breathed in the fine fibers shed by the spindles and looms. Metal grinders, masons, pottery-factory workers—all were routinely exposed to occupational dusts that kept their lungs in a perpetual state of low-level inflammation. Wider still were the effects of coal- and wood-burning. In cold weather, many cities, especially in river valleys, were enveloped in a haze of particulates from the ubiquitous coal fires. Small-town and rural dwellers constantly inhaled wood smoke during the winter. It is little wonder that observers of the national scene in the decades before the Civil War spoke routinely of consumption and phthisis as ubiquitous, and how it accounted for one of five, or one of seven adult deaths.[27]

Ironically, the dramatic symptoms of active tuberculosis infection arise not from the rather indolent and unaggressive bacillus itself, but from the immune system's vigorous attempts to oust it. After initial exposure, or upon reactivation, a delayed hypersensitivity reaction floods the area of a tubercle with inflammatory molecules. The lesion enlarges, liquefies at the center, and then breaks down. This opens up cavities in the lungs where the tubercle bacilli multiply, invading and destroying the solid bronchi and blood vessels. From there they can easily spread to the outside when the patient coughs and talks, and the disease process starts again.

Nineteenth-century American physicians, like their forebears going back to the Greeks, were intimately familiar with the classic symptoms of active pulmonary consumption: low-grade fever, night sweats, chills, fatigue, general malaise, loss of appetite and weight, irregular menses in women, and finally the characteristic cough. Paroxysms racked the patient as he or she tried to bring up fluid and dead lung tissue. These rose reluctantly in a dense, viscous sputum—"ropy" was the most common way to describe it. If a small blood vessel broke, the sputum would be blood-tinged. John Keats, who was medically trained, instantly recognized his fate when he saw his bloodstained pillow in February 1820; he died a year later.[28] If a larger vessel eroded, coughing could bring on an uncontrollable hemorrhage and inevitable, quick death. For many, this death was preferable to the usual lingering one, in which the patient lay in bed, surrounded by family, and slowly lost weight, energy, and respiratory capacity. Weakened, feverish, breathing ineffectually in a close room, the victim would finally expire.

Because of the chronic and often remitting nature of consumption, the long course of the disease provided many opportunities for the anxious sufferer to feel better, and to believe temporarily that he or she was improving. As with cancer today, the protracted course also gave ample scope to try all kinds of medical and lifestyle remedies. Physicians responded with treatments and advice. In 1882 Frank Woodbury, a leading practitioner in Philadelphia, prescribed codeine for cough, atomized limewater and belladonna for inflamed lungs, gargles of lemonade and egg whites for irritated larynxes, and oxygen and morphine for air hunger. For extra-pulmonary care he recommended digitalis for palpitations of the heart, a salicylic for fever, atropine for night sweats, and small quantities of strychnine as a tonic to the nervous system. The patient's diet was to be palatable and easily digested, and was to include one to three eggs a day. Good ventilation and fresh air were an absolute necessity, as were loose clothes that were regularly aired, attention to skin hygiene, and a good horsehair and spring mattress.[29] With so many interventions projected upon the waxing and waning of the disease, a well-heeled consumptive would have numerous occasions to thank his doctor for efficacious care and pay out large sums of money. In nineteenth-century America, this was justified because medicine had a highly developed set of ideas about the cause and cure of pulmonary tuberculosis.

Changing Ideas of Consumption, 1850–1882

At midcentury, European and American physicians could feel rather satisfied that phthisis of the lungs, popularly known as "consumption," was well understood. Several generations of English, German, and especially French clinician/pathologists had performed tens of thousands of autopsies on phthisical patients and agreed that the disease had clear anatomical features. The tiny tubercles grew in size, led to the formation of cavities, and then brought on the symptoms of cough, fever, wasting, and loss of breath, and eventually death. Similar disease processes were found in other organs, such as the brain and its coverings (the meninges), bones and spine (as with Pott's disease), lymph glands (manifested as scrofula), intestines, kidneys, bladder, thyroid, testicles, and uterus—all characterized, more or less, by the same tubercles that were found in the lungs. A growing number of scientific physicians were beginning to agree with the idea, apparently first suggested by Jean Nicholas Papavoine in 1830, that there was a single disease, tuberculosis, and that it could manifest itself in various parts of the body.[30] Researchers were also beginning to examine tubercles under the newly improved microscopes, which usually tended to support this unitary concept of the disease.[31] Indeed, the name "pulmonary tuberculosis" rapidly gained ground in the medical profession in the 1850s and 1860s and became interchangeable with the older term, "phthisis pulmonalis." After René Laennec invented the stethoscope in Paris in 1816, he and his followers in many lands worked out excellent techniques for diagnosing tuberculous lungs.[32] Numerous physicians, like one in Buffalo in 1858, cast scorn on the depletive therapies—bleeding, purging, vomiting, blistering—so popular twenty years before. There had been a "complete revolution" in the management of consumption. Physicians now relied, he said, not on "any special curative agent," but on conservative treatments focused on hygiene, alleviation of symptoms, a nutritious diet, and ventilation.[33]

Unlike smallpox and measles, which were obviously contagious, or cholera, which John Snow had recently shown was transmitted via water contaminated by human waste, consumption was held in most parts of Europe and the United States to be a "constitutional" disease. It arose from the interaction of hereditary predisposition, individual behavior, and local environment. It existed on a very wide spectrum; it was a fundamental mode of the human body's interaction with the world. Layman and physician alike agreed that "everyone is a little bit consumptive."

Hereditary predisposition was clearly important.[34] How else, physicians reasoned, could one account for the appearance of the disease in one or both parents, then in some of the children, and thereafter even in some of the third generation? One particularly famous case was the family of Ralph Waldo Emerson. His father was the first to succumb, in 1811, at the age of forty-two. Thereafter Ralph, as a young student at Harvard in the 1820s, developed consumption, and it dogged him for the rest of his life. Three of his brothers came down with the disease, two of them dying of "galloping consumption" in their twenties. Ralph's first wife, Ellen, had consumption when he married her in 1829, and she did not last two years. He married again, and tuberculosis struck among his children and grandchildren. Similarly, Henry David Thoreau's grandfather, father, and sister all died of consumption, as did he in 1862, at the age of forty-five.[35] The very capriciousness of the disease's incidence, however, argued against any ironbound necessity. Some individuals in practically all families escaped the taint.

Standard medical opinion was that heredity expressed itself in a consumptive "diathesis," a bodily constitution that was predisposed to a particular disease—in this case, tuberculosis. Physicians claimed that they could easily recognize the consumptive diathesis, especially in the characteristics of women who were tall, erect, and slender, and who had white, translucent skin, bright eyes, fine and silky hair, and little physical strength or stamina. The notion of such a phthisical habitus went back to Hippocrates but had been especially developed by medical and popular writers alike in the first half of the nineteenth century.

While heredity constituted a predisposing cause, behavior could precipitate the disease. The anecdotal literature is full of instances in which exposure to cold or wet weather was seen to bring on an attack.

Drafts were dangerous. Failure to treat bronchitis could cause it to develop into phthisis. Failure to eat heartily only pushed the tubercular body further down the path to an active case. Consumption was common among the immigrant classes, especially the Irish, it was thought, because of their alcoholism, immoral habits, and lack of personal hygiene.

The final piece in the etiology puzzle was the local environment. Place mattered. If there was any proposition that would have gained almost universal assent around 1860, from doctors as well as from the general public, it was that the patient's immediate surroundings were intimately connected with health and disease. This belief was built on a tradition of medical concern with locality that went back to Hippocrates' *Airs, Waters, and Places*, and that was only strengthened by the miasmatic theory of epidemic disease, which held almost unquestioned sway in the first half of the nineteenth century. The country was healthier than the city. Well-drained highlands caused less disease than damp or swampy lowlands. Clean air caused less lung irritation and general debility than foggy or polluted air. Sunshine was good, as were clear skies. The locale in which you lived could, literally, be the death of you.

Beginning in the 1850s or so, however, this medical environmentalism underwent a subtle shift. Local surroundings still mattered, but region and climate, rather than locality and weather, rapidly took on greater importance. This was especially true in the United States, where mobile patients could move from coastal marshlands to alpine woods, from bracing Maine to humid Florida, from cold and blustery Chicago to the newly acquired territories of New Mexico and Arizona. In 1860 a California physician, John Blake, entered into the medical literature one of California's first claims for the treatment of pulmonary consumption.[36] Thoreau spent the long summer of 1861 in Minnesota, a state that likewise put forward its claims for salubrity.[37] Aiken, in the hill country of South Carolina, became a winter retreat for northerners.[38] The Gulf Coast of Florida argued its merits.[39] By the early 1870s there was an active discussion in the medical literature about the regions and towns where the suffering consumptive could go to recover his health. In general, opinion held that the pulmonary patient had to move out of the north and northeast, toward the west and south, to drier, sunnier climates, and possibly higher altitudes.

Acting on such beliefs was made all the more possible by the rapid westward expansion of railroads in the 1860s and 1870s. Colorado was an early beneficiary. Denver had only about 4,800 residents in 1870, and it boomed to 35,600 in 1880, becoming number fifty in the rank of American cities. By 1890 it had tripled to 106,700, becoming the twenty-sixth-largest American city—the approximate rank it has occupied ever since. Colorado Springs, at an elevation of more than six thousand feet, became even more specialized in the care of consumptives during its mild and dry May to October season.[40] Farther west, the transcontinental railroad linked San Francisco to the east in 1868, and the Southern Pacific extended its rails south to Los Angeles by 1876. Within a few years Southern California was also easily reachable directly via the southern route of the Atchison, Topeka, and Santa Fe. Consumptives could now move rapidly and comfortably to all those climates that medical opinion had come to recognize as curative.

Latitude, climate, and topography were not just restorative; they were seen as preventative. Edmund Andrews, a Chicago physician and prolific writer on a broad range of medical topics, reflected that belief in a preliminary article in 1866,[41] and more fully in one on "Climatic Relations of Cancer and Consumption" in 1885. He compared the mortality schedules from the 1870 and 1880 U.S. censuses to conclude that, whereas cancer seemed to be on the rise, the incidence of consumption was declining. The "white plague" was still, however, distributed geographically in a way that was alarming. In the table shown on the next page, I extract some of his reworking of John Shaw Billings' mortality analysis:[42]

Region	Consumption Deaths to All Deaths
New England seacoast	1 to 6
Middle Atlantic seacoast, N.Y. to Va.	1 to 7
Prairie region of Ia., Ill., Kans., Neb., Wis., Minn., etc.	1 to 10
South Atlantic seacoast, N.C. to Ga.	1 to 11
Southern interior plateau, S.C., Ga., etc.	1 to 12
Western plains, Dak., Mont., Wyo., Colo., N.M., etc.	1 to 14
Southwest central region of west Tex., etc.	1 to 15

Andrews concluded that consumption rates decreased as one went south, away from the seacoast, into higher, drier, and then more western land. The extreme was reached in New Mexico, in which only one out of fifty deaths was caused by consumption.

Thus, at the time Baur takes up his Southern California narrative, around 1870, medical beliefs about the power of region and climate, little organized and rather off-hand three decades before, had cohered into a powerful set of imperatives. Not only should consumptives forsake the climate in which the disease struck, but they also needed to move to a place where their children could live in safety. This clear picture of the relationship between consumption and region, however, became muddled just as it approached perfect resolution.

Koch, Bacilli, and the Contagiousness of Tuberculosis

European and American medical practitioners were stunned on March 24, 1882, when Robert Koch announced in Berlin that he had discovered that tuberculosis was caused by a bacillus.[43] The bacterium was small, difficult to culture, and even more difficult to stain and see, but it was there. Koch had isolated bacilli from tubercles located in many parts of patients' dead bodies. He had cultured the bacteria in a liquid medium according to a protocol of his own design. He had injected small samples of the cultured bacilli into a uniquely susceptible experimental animal, the South American guinea pig—until then, known largely as an exotic pet. The animals contracted tuberculosis. From them, he could then re-isolate the bacilli that had proliferated in causing the disease. This set of protocols—isolating, culturing, infecting, re-isolating—which soon became known as "Koch's postulates," has served as the gold standard for proving disease causation ever since. Koch's coup in explaining the cause of tuberculosis was all the more credible because it had been accomplished by a master. At a time when Germany led the world in scientific medicine, he was the leading figure in the German Imperial Health Office. Since 1876 he had published paper after paper laying out the major principles of the germ theory of disease, thereby putting a firm laboratory foundation under Joseph Lister's 1868 announcement of the techniques of antiseptic surgery. Italian and Spanish folk beliefs had long held consumption to be infectious. French, and especially German, experiments in the 1860s and 1870s had shown that diseased tissue could infect another animal. Koch was alone in not only finding the bacillus but also in developing the panoply of procedures to visualize it and prove its pathogenic nature.[44]

Word spread rapidly in central Europe and beyond. The *Boston Medical and Surgical Journal*, edited by doctors who had done their advanced training in Germany and Austria-Hungary, published a tantalizing abstract, "The Bacillus of Tuberculosis," on April 20, 1882.[45] *The Lancet* of London must have had a correspondent on the spot, because it published a full account of Koch's lecture on April 22, just a few days after his preliminary report appeared in print. The *Lancet* article emphasized the precise and detailed laboratory procedures necessary to find the bacilli, and noted that Paul Clemens von Baumgarten, a pathologist at the University of Königsberg, had independently confirmed Koch's discovery. The weekly ended its article with these prophetic words:

> The pathological importance of the discovery of the proximate cause of this frightful scourge of the human race cannot be over-estimated, nor is it possible to foretell the practical results to which it may lead.[46]

Within a few months Koch's discovery was the talk of the American medical journals. They reported that Koch had demonstrated his bacilli on April 20 at the German Congress for Internal Medicine at Wiesbaden, which started a vigorous discussion of how to interpret the findings.[47] By May 25, 1882, William Whitney, a German-trained Harvard pathologist, had duplicated Koch's procedures and come up with the same results.[48] European medical correspondents, as well as Americans traveling in Europe, wrote letters back about the great debate that ensued.[49] One wrote to Chicago from Vienna in July: "The tubercle bacillus is still the lion of the day, and bacillus hunting the chief occupation of all interested in pathology."[50] Everyone acknowledged that the organisms were present in tuberculous tissue and tuberculous sputum, and work was proceeding apace to prove that all those forms of tuberculosis that had been painstakingly forged into a unitary theory of the disease were in fact derived from the common presence of the tubercle bacillus. During the rest of 1882–83 there were some Americans, such as the microscopic pathologist Henry Formad in Philadelphia and the pathologist Henry Schmidt of New Orleans, who could not find the bacilli in diagnosed cases of tuberculosis.[51] Even though Formad took the opportunity to visit Koch's Berlin laboratory in 1883 to learn firsthand the techniques of culturing and staining, he and Schmidt remained unconvinced of a proved causal link between the presence of the tubercle bacillus and the disease.[52] Medicine simply left them behind, and by 1885 or 1886, virtually all of the leaders of American medicine acknowledged that phthisis was caused by growth in the lungs of the tubercle bacillus. Like wound infections, gonorrhea, pneumonia, meningitis, cholera, diphtheria, and anthrax, tuberculosis was yet another disease that had met its match in the laboratory. Consumption was an infectious disease.

The problem that remained unsolved through the 1880s and early 1890s was that the newly emergent bacterial disease of "pulmonary tuberculosis" didn't seem to act like an infectious disease. As thousands of physicians worldwide learned to culture the tubercle bacillus and identify it under the microscope, they also learned that it could be detected in many cases in which the patient did not have overt disease, or even the characteristic lung tubercles and cavitations found in the autopsies of consumptives. The bacillus could be discovered and cultured in the remains of sputum, even in the dust in the sickroom. It seemed that the old proverb "everyone is a little bit consumptive" needed to be modified to read: "Everyone is exposed to, and perhaps infected with, the tubercle bacillus." But not all developed the disease.

It was here that the older conceptions of inherited and environmental disease could come into play. Clearly what made a difference was an inherited weakness and inclination toward developing the active disease, coupled with an environment (climate, in particular) that aggravated rather than ameliorated it. A new tripartite balance was conceived, with the three elements being: the infective (but ever-present) agent, a hereditary predisposition, and the precipitating environment. Since the ubiquity of the tubercle bacillus seemed beyond remedy, and heredity could certainly not be changed, physicians could still think of the environment as the primary tool that they had at their disposal to retard, and possibly cure, the disease.

Koch's even greater bombshell, his announcement in August 1890 of a putative therapy for tuberculosis, tipped the balance decisively toward reconceptualizing consumption as an infectious disease.[53] Hundreds of European and American doctors streamed to Berlin to learn firsthand about the cure, "tuberculin." The arrival of minute supplies of "Koch's lymph" was tracked breathlessly in the newspapers. Everyone wanted to get some. Moreover, tuberculin fit nicely into the stream of laboratory-created vaccines and therapies that emerged between 1887 and 1895, of which the greatest triumph was diphtheria antitoxin, announced in 1892 by Koch's younger colleague, Emil Behring. Even as it became clear in the first half of the

1890s that tuberculin was nowhere near the sovereign remedy that Koch had hoped for, it was modified, purified, and amplified in such a way that it retained a central place in the profusion of "consumption cures" that were announced in that decade. The discovery of X-ray procedures in 1895, and their rapid deployment thereafter as possible tools—though not yet widely used—to confirm a diagnosis of pulmonary tuberculosis, only seemed to bring tuberculosis more fully into the realm of ordinary disease. Medicine treated it by identifying, controlling, and destroying the pathogen.

Disinfection, Registration, and Separation into the Sanatorium

Before the late 1880s, few laymen or physicians felt that a consumptive was any kind of danger to the community. They coughed and spat, and sat wanly in their rooms or on the porches of health resorts. Southern California promoters actively sought the less sickly of such well-heeled visitors or new settlers. New health migrants brought money, skills, and culture, and might recover to become model citizens. If sicker consumptives died friendless and penniless, their burial became a regrettable burden on the community. But both kinds were often objects of muted sympathy, a reminder of the unfathomable grace that had spared the onlooker. By the early 1890s, however, this benign engagement was beginning to be transformed by a deepening public understanding of how tuberculosis spread. Consumptives slowly changed from pitiable souls to pitiable potential dangers.

Sputum was only the first target of disinfecting frenzy. As early as 1889, Hermann Biggs, the chief medical officer of the New York City Board of Health, summarized the evidence that sputum dried into dust that could communicate tuberculosis when injected into animals.[54] This dust was a danger. Municipalities, not least in health-seeker regions, enacted ordinances against spitting on the sidewalk or floor. Spittoons were mandated in public places. Rooms, furniture, and clothing came next. If a consumptive died in a hotel or boarding house, local regulations required a chemical scrub-down and the burning of any material or objects that couldn't be disinfected by boiling. In the 1890s physicians gradually came to realize, moreover, that spitting and fomites (objects in contact with an infected person) were much less dangerous than coughing. Bacteriologists had done simple experiments to show that a cough, or much more a sneeze, could spread droplets of bacilli twelve to twenty feet. Family members, or even visitors in the same room, were potential victims of infection. Anyone who sat next to a consumptive at dinner, or across from one at a card table, or who even passed one in the street, was exposed to some level of danger.

Such a threat to public health could only be countered, officials like Biggs argued, if the authorities knew which persons had active cases of pulmonary tuberculosis. You can't control a disease if you don't know who has it. In 1894, over the objections of many private physicians who argued for a lesser contagiousness of tuberculosis, New York City made consumption a reportable disease. The city was in advance of many others, but notification only built momentum thereafter.[55]

One might suppose that it was an easy next step to create facilities that separated tuberculars, such as that emerging institution, the sanatorium.[56] It had, however, more complex roots than as a place for quarantine for consumptives. The first sanatorium was established in 1854 at Görbersdorf (now Sokołowsko, Poland) in the mountains of Lower Silesia, in the eastern part of Germany near the Czech border, at an altitude of two thousand feet. Its founder, Dr. Hermann Brehmer, had cured his own phthisis during an extended sojourn in the Himalayas. He believed passionately that fresh, cold air, sunshine, directed exercise, frequent, nourishing meals, and a regimented schedule would have a healing effect on his consumptive patients. Experience bore him out. During the 1860s and 1870s similar private sanatoriums were organized elsewhere in rural German-speaking areas; the most famous was at Davos, in eastern Switzerland, where Robert Louis Stevenson spent the winter of 1880. The ideal of a health institution in a healthy place came to America first at Saranac Lake, in upstate New York. The consumptive physician Edward Livingston Trudeau arrived there in 1876 to treat his pulmonary symptoms with an outdoor life. He read about Brehmer's sanatorium and decided in 1882 to organize a similar one on a smaller scale: the Adirondack Cottage Sanatorium. It soon became successful and acted as a beacon of hope, especially for early-stage consumptives.[57] Thereafter the sanatorium movement in America spread slowly at first, but by the mid-1890s gathered pace. As befits a health destination, Southern California had early small representatives of the species, such as La Viña Sanatorium in the foothills north of Pasadena.

The sanatorium, an explicitly medical facility aimed at the long-term care and cure specifically of tuberculosis patients, grafted itself onto the older notion of the sanitarium. As used in the 1860s, 1870s, and 1880s, the term "sanitarium" meant a kind of locale where visitors gathered to improve their health. It could be a hotel, or even a seasonal tent-camp—such as several in the San Gabriel Mountains near Los Angeles in the late 1880s—where residents with all kinds of ailments came to relax, mingle, and recover from the strenuous capitalist life of the Gilded Age. The theme was one of outdoor life, made possible year-round in a benign climate like that of Southern California.[58] The sanitarium embraced not only consumptives but also asthmatics, anorexics, neurasthenics, and simply the wealthy bored found at any destination spa. It could even be a site or town where residence was seen to confer health benefits. William Chamberlain, a wealthy New York City physician and consumptive, expressed the idea well in the title of his 1887 article, "Pasadena as a Sanitarium."[59]

By the late 1890s, at the end of Baur's time window, this older notion of health-giving place was morphing into one of curative institution. Growing medical—and popular—knowledge of the infectious nature of tuberculosis made it distinctly less desirable for Southern California residents or promoters to welcome consumptives, especially very sick ones, as new residents. Locals still broadcast the message of the healthfulness of Southern California, and physicians agreed with them. But they also emphasized that for the seriously ill, the local climate and lifestyle could be safely utilized only if the migrant lived in the much more regimented environment of a sanatorium—hence the rapid proliferation of institutions represented by the Barlow, Pottenger, and Los Angeles sanatoriums. Health seeking in Southern California had been transformed by the contemporaneous revolution in the medical understanding of tuberculosis.

Postscript 2009:
New Meanings for Tuberculosis, Climate, and Lifestyle

In their heyday, from around 1900 through the 1940s, sanatoriums were universally recognized as the best possible means of dealing with the tuberculosis threat. Patients with active disease could be identified not only clinically but also by the characteristic bacilli in their sputum. They could be confined, treated with an open-air life and good nutrition, and then released back into society. Multiple studies demonstrated that such "arrested" cases were no longer infectious—as our good Dr. Brady wrote in the *Los Angeles Times*. Physicians could feel a sense of triumph as the rates of tuberculosis cases and deaths plunged.

With the coming of antibiotics and chemotherapy, however, the medical attitude toward environmental treatment changed. First came streptomycin and para-aminosalicylic acid in the 1940s, then isoniazid and pyrazinamide in the 1950s, and rifampin in the late 1960s. By that time, tuberculosis specialists focused on developing exactly the right combination to wipe out the bacilli—even if it did require two to fifteen months of daily medications that had nasty side effects. The belief that tuberculosis was caused or cured by climate and lifestyle was pooh-poohed as a relic of a prescientific age.

Ironically, it is the "sunshine" vitamin D that has prompted a very recent reevaluation of how the environment relates to tuberculosis. In the early 1920s British and American medical researchers discovered a substance (dubbed "vitamin D" because of its place in the sequence of vitamins identified) that is created when sunlight strikes mammalian skin, and which is absolutely necessary for the growth, mineralization, and maintenance of bones. A vitamin D deficiency in children produces rickets, which causes bone and joint malformations. The disease was rampant in the large northern cities of nineteenth-century Europe and America. A deficiency in adults produces osteomalacia, characterized by generalized aches and frequent falls and fractures. In the 1930s the animal form created in the skin, vitamin D3 (cholecalciferol), was isolated, and thereafter synthesized in the laboratory. In the postwar period it was made a mandatory fortification for milk, and rickets essentially disappeared in the developed world. Scientists recognized that persons with darker skin, such as African-Americans, needed much more sun exposure to produce the necessary amount of vitamin D, but it was felt that most individuals had a sufficient level. For decades thereafter, the vitamin had a small but established place in human biology and medicine.

Not until the late 1980s, as researchers developed the laboratory tools to track its paths in the body more closely, did vitamin D start to appear very much more interesting.[60] From the vitamin the liver and kidneys produce a hormone-like substance (calcitriol) that is distributed via the blood to major systems and organs: the intestines, pancreas, thyroid and parathyroid glands, bones, skin, lungs, heart, muscles, and brain. All these parts of the body have receptors that are activated beneficially by the hormone. By 2006 ample vitamin D levels had been implicated in preventing certain cancers and promoting recovery from them, moderating high blood pressure and reducing the risk of a heart attack, alleviating autoimmune diseases such as diabetes and multiple sclerosis, and modulating the immune system to protect against bacterial and viral invasion.

Among the most important diseases affected by vitamin D is tuberculosis. More than twenty years ago a British lung specialist proposed that vitamin D deficiency among immigrants to Britain from the Indian subcontinent was responsible for the extremely high (thirtyfold) incidence of tuberculosis among them. From bits of emerging evidence, he presciently suggested that this occurred because a lack of vitamin D impaired the immune response, especially that of the macrophages.[61] Only since the late 1990s has this idea been pursued vigorously. There are now scores of laboratory, clinical, and epidemiological studies that detail how vitamin D enhances immunity, especially how it induces macrophages to produce great quantities of a molecule necessary to kill resistant tubercle bacilli once the immune cell has engulfed them. With low levels of vitamin D, the bacterium has a much greater chance of destroying its would-be destroyer. To win, the body needs much higher levels of vitamin D than those that are merely adequate to grow and maintain bones.[62]

Over the same recent decades, scientists have come to understand the detailed mechanisms by which sunlight produces vitamin D in the skin. Only a narrow band of invisible wavelengths (295–300 nanometers) in the ultraviolet B (UVB) spectrum can synthesize the vitamin optimally.[63] This band becomes greatly attenuated when passing through the earth's atmosphere at an angle, so UVB can only be active when the sun is high in the sky. Since the maximum height of the winter sun is reduced as latitude increases, penetration of UVB drops off much more than simply because of the shortness of the winter days.[64] The inhabitants of cities such as Boston, Providence, Hartford, New York, Cleveland, Detroit, Chicago, Milwaukee, or Minneapolis have so little incident UVB that, even if they sunbathed in the snow, they would not be able to synthesize *any* vitamin D between October and February. Residents of cities even farther north, such as London, Paris, Brussels, Amsterdam, Cologne, Berlin, Warsaw—not to speak of Scotland or the Scandinavian capitals—are similarly precluded from November through March.[65] Humans in these latitudes must subsist on vitamin D stores from the summer, which decline rapidly. Even during seasons that do have UVB incidence, many city dwellers have their exposure limited by indoor occupations and narrow streets hedged closely with multistory buildings. In addition, UVB is filtered out by clouds, pollution, and especially by fog, either from natural causes or from coal burning. UVB is incapable of penetrating ordinary glass windows. Even when available, UVB can't synthesize the vitamin when the person's head, trunk, arms, legs, and feet are covered with hats, clothes, and shoes.

Historians will never be able to measure the vitamin D levels of middle-class Americans in the cities of the Gilded Age, but they must have been dire. High latitudes, cloudy weather, urban pollution, indoor occupations, dense cities built to the sidewalk, clothing that covered the body entirely, few customs of outdoor recreation, especially for women—all of these factors would have conspired to make people unhealthy in fundamental ways.

Seen in this (sun)light, we can understand why those who were newly arrived in late-nineteenth-century Southern California, whether healthy or health seeking, would have felt such a surge of well-being. Los Angeles and San Diego lie on the same latitude as Casablanca and Marrakech. The sun was high, the sky often cloudless. Smoke and pollution were not present. Rainstorms were seldom and brief. Winter nighttime lows could be chill, but the daytime temperatures enticed residents into an outdoor life. Indeed, that was often why the migrant came to Southern California. Riding, bicycling, hiking, camping, even just walking—all done for the "open air"—brought sun exposure. Garments covered less, and hats were smaller or nonexistent. Even if a consumptive only spent a few hours lounging in the shade on a veranda, he or she could absorb a more-than-adequate dose of indirect UVB.[66] In Southern California it was not the bracing dry air that cured, but the sun.

Considered in this way, Baur's *Health Seekers* is more than an exquisitely executed study in social history, popular attitudes, and forces behind urban growth. It is a chronicle in miniature of a remarkable human experiment in which hundreds of thousands of migrants unwittingly medicated themselves. In the process, many of the health seekers found what they sought.

Further Reading

This bibliography of selected books published during the last three decades includes only those that are widely available in public, as well as college, libraries. Almost all can be purchased from online used booksellers.

Health, Migrants, and the American West

Abel, Emily K. *Tuberculosis and the Politics of Exclusion: A History of Public Health and Migration to Los Angeles.* New Brunswick, N.J.: Rutgers University Press, 2007.

Picking up where Baur leaves off, at about the year 1900, Abel skillfully dissects the way in which stigmatized groups of migrants, especially Mexicans, Filipinos, and African-Americans, were herded into paternalistic "treatment" programs and sanatoriums, and then, with the onset of the Depression, were targeted by public agitation and policies that excluded them from society.

Abel, Emily K. *Suffering in the Land of Sunshine: A Los Angeles Illness Narrative.* New Brunswick, N.J.: Rutgers University Press, 2006.

Abel has based this perceptive, intimate, and tautly written biography on the splendid collection of the letters of Charles Dwight Willard (1866–1914) at the Huntington Library. Willard moved to Southern California as an invalid in 1886, recovered enough to work as a journalist, and spent the rest of his life battling his tuberculosis and boosting his adopted city.

Jones, Billy M. *Health-Seekers in the Southwest, 1817–1900.* Norman, Okla.: University of Oklahoma Press, 1967.

Jones covers a long time period and a wide geographical area, including Texas, New Mexico, Arizona, Colorado, Utah, and Nevada; for California, he largely refers the reader to Baur. Although much less rich in social history than Baur's volume, Jones's study provides interesting discussions of perceived indigenous diseases of the West, the medical specialty of climatology as it related to consumption, Colorado resorts, and the demise around 1900 of the legend of the health-giving West.

Tuberculosis and Society

Dormandy, Thomas. *The White Death: A History of Tuberculosis.* London: Hambledon Press, 1999. Reprint, New York: New York University Press, 2000.

Dormandy provides the best multifaceted history of tuberculosis for the period beginning about 1800 through to the mid-1950s. In a predominantly Euro-centered discussion, he touches on every aspect of the disease: notions of cause, clinical diagnosis, the disease's relationship to poverty, climatic cures, Koch's discovery of the bacillus, tuberculin and X-rays, the sanatorium from Switzerland to the United States, surgical therapies such as pneumothorax and thoracoplasty, vaccines and BCG, and quack remedies. Interspersed are numerous evocative vignettes about famous patients from Keats to Kafka, including Napoleon II, Robert Louis Stevenson, and Anton Chekhov.

Daniel, Thomas M. *Captain of Death: The Story of Tuberculosis.* Rochester, N.Y.: University of Rochester Press, 1997.

Daniel's loose mixture of historical discoveries and explanations of contemporary medical concepts is organized around four themes: tuberculosis through the ages, the infectious nature of the disease, susceptibility and resistance, and treatment.

Ott, Katherine. *Fevered Lives: Tuberculosis in American Culture since 1870.* Cambridge, Mass.: Harvard University Press, 1996.

Ott's well-written study focuses on the period from about 1870 to 1930, emphasizing popular and literary perceptions of the disease, the material culture of the sickroom, the patient's labor to get well, government surveys of the urban ecology of the sickly, and psychological patterns of life in sanatoriums.

Feldberg, Georgina D. *Disease and Class: Tuberculosis and the Shaping of Modern North American Society.* New Brunswick, N.J.: Rutgers University Press, 1995.

Dealing with both the United States and Canada, Feldberg traces the role of tuberculosis in society before Koch's discovery, the reception of bacterial etiology, and the campaigns between 1900 and 1925 to educate the populace and build resistance to the disease. The latter half of the book focuses on the contention in North America over using BCG as a vaccine.

Rothman, Sheila M. *Living in the Shadow of Death: Tuberculosis and the Social Experience of Illness in American History.* New York: Basic Books, 1994.

The experience of the patient is virtually the exclusive focus of this beautifully written reconstruction, built up from the letters and diaries of 110 patients. The later chapters on sanatorium life are excellent, but the volume's center of gravity is the period from about 1810 to 1890, including insightful discussions of narratives written by those who heeded the call to "come west and live."

Ellison, David L. *Healing Tuberculosis in the Woods: Medicine and Science at the End of the Nineteenth Century.* Westport, Conn.: Greenwood Press, 1994.

This well-researched assessment of the life of Edward Livingston Trudeau (1848–1915) interweaves his own experience as a consumptive, his research on the tubercle bacillus and tuberculin, and his leadership at the Adirondack sanatorium at Saranac Lake. Especially useful is Ellison's discussion of the transition between 1880 and 1895 from the hereditary view of tuberculosis to the bacterial view.

Ryan, Frank. *The Forgotten Plague: How the Battle against Tuberculosis was Won—and Lost.* Boston: Little, Brown and Co., 1993.

Ryan provides a multistranded and historically rich narrative of the search for anti-tuberculosis chemotherapies such as streptomycin, PAS, and isoniazid, covering the late 1920s to the mid-1950s. From primary materials, Ryan weaves together the interacting research of Selman Waksman in the United States, Gerhard Domagk in Germany, and Jörgen Lehman in Sweden to produce a lively and very readable story.

Bates, Barbara. *Bargaining for Life: A Social History of Tuberculosis, 1876–1938.* Philadelphia: University of Pennsylvania Press, 1992.

Bates is a distinguished clinician whose textbook has taught generations of American medical students the subtle art of examining their patients. She turns an equally discerning eye on 169 volumes of the correspondence and papers of Lawrence Flick (1856–1938), the leading tuberculosis expert of his generation, to draw a fascinating portrait of doctors and their patients in Philadelphia and eastern Pennsylvania.

Caldwell, Mark. *The Last Crusade: The War on Consumption, 1862–1954.* New York: Atheneum, 1988.

This is an evocative exploration by a literary critic of the images and emotional valences of the sanatorium movement in America, from Trudeau's work in the Adirondacks in the 1880s to the closing of the sanatoriums in the 1950s, with emphasis on routines, rituals, and implicit cultural values.

Teller, Michael E. *The Tuberculosis Movement: A Public Health Campaign in the Progressive Era.* New York: Greenwood Press, 1988.

This volume covers American public health organizations and their programs from the 1880s to 1917.

Smith, Francis Barrymore. *The Retreat of Tuberculosis, 1850–1950.* London and New York: Croom Helm, 1988.

This detailed study of the decline of tuberculosis in Britain focuses not only on the usual elements (statistical benchmarks, changing ideas of cause, attempts at individual treatment, dispensaries and sanatoriums, the relationship of the disease to occupational hazards and poverty, and possible vaccines) but also on the often-obstructive attitudes and actions of the medical profession there. No similar analysis exists for the United States.

Taylor, Robert. *Saranac: America's Magic Mountain.* Boston: Houghton Mifflin, 1986.

Taylor provides an anecdotal history of the famous figures who sought revitalization of their health in the Saranac Lake region of the Adirondacks, from Ralph Waldo Emerson to Sylvia Plath.

Keers, Robert Young. *Pulmonary Tuberculosis: A Journey down the Centuries.* London: Baillière Tindall, 1978.

This well-written and accessible survey by a scholarly physician and tuberculosis expert looks at the changing medical and scientific understanding of pulmonary tuberculosis and its treatment. The latter two-thirds of the book covers the period from about 1890 to 1970.

Notes

1. "Teamwork," *Los Angeles Times*, November 5, 1947, A4.
2. "Nurses Critically Needed by Most Hospitals in Southern California," *Los Angeles Times*, December 14, 1947, A1–A2.
3. William Brady, "Here's to Health: Questions and Answers: Tuberculosis," *Los Angeles Times*, March 6, 1947, A7.
4. William Brady, "Here's to Health: Compulsory Vaccination," *Los Angeles Times*, November 17, 1947, A9.
5. William Brady, "Here's to Health: So Your Tuberculosis is Not Active Right Now?" *Los Angeles Times*, March 13, 1947, A7.
6. William Brady, "Here's to Health: Persons, Not Things, Spread Germs," *Los Angeles Times*, June 23, 1947, A8.
7. "The Governor's Inaugural Address," *Los Angeles Times*, January 7, 1947, A4.
8. "Public Health Fight Urged. Doctors Told They Must Meet Danger of State Medicine," *Los Angeles Times*, May 1, 1947, A1.
9. "Portable X-Ray Tests Workers," *Los Angeles Times*, January 29, 1947, 2; "X-Ray Tests Offered at 'Times' Boys Club," *Los Angeles Times*, May 6, 1947, A12.
10. "X-Ray Checkup of 2000 at U.C.L.A. Slated," *Los Angeles Times*, February 17, 1947, 12.
11. "Teachers and Other School Employees Take X-Ray Tests," *Los Angeles Times*, December 2, 1947, A1; Elinor Gene, "Food for Thought: What's New in the PTA?" *Los Angeles Times*, June 17, 1947, A6.
12. These occurred very frequently through the year, for example, *Los Angeles Times*, March 30, 1947, F27.
13. "Stars to Appear at Duarte Benefit," *Los Angeles Times*, March 30, 1947, A3.
14. Lee Shippey, "Leeside," *Los Angeles Times*, May 2, 1947, A4.
15. "Hollywood C. C. Women's Tea to Honor Volunteers," *Los Angeles Times*, December 5, 1947, A5; "Southland G.O.P. Women Arrange Gift Luncheon," *Los Angeles Times*, December 14, 1947, G6.
16. The classic history is Richard H. Shryock, *The National Tuberculosis Association, 1904–1954: A Study of the Voluntary Health Movement in the United States* (New York: National Tuberculosis Association, 1957).
17. "Hollywood C. C. Women's Tea to Honor Volunteers," *Los Angeles Times*, December 5, 1947, A5.
18. See *Los Angeles Times*, November 24, 25, 26; December 4, 5, 8, 15, 26, 1947.
19. "New Drug Hailed in Fight on Tuberculosis," *Los Angeles Times*, July 5, 1946, 2.
20. "Parents Gamble on Ill Son's Life," *Los Angeles Times*, September 8, 1946, 15; "Teamsters Widen Tie-Up of Drug Deliveries Here," *Los Angeles Times*, September 21, 1946, A1; "New Drug Going in Trade Channels," *Los Angeles Times*, November 30, 1946, 3.

21. "Streptomycin Injected in Effort to Aid Child," *Los Angeles Times*, November 15, 1946, A1, 3; "New Peril Enters Baby's Life Fight," *Los Angeles Times*, November 19, 1946, A1.
22. "Bold Test of Miracle Drug Saves Baby's Life," *Los Angeles Times*, January 11, 1947, A1.
23. "Doctor to Discuss 'Miracle Drugs,'" *Los Angeles Times*, March 21, 1947, 5; an address at the Los Angeles Sanatorium by Bogen on streptomycin is noticed in "Sanatorium Staff Will Meet Today," *Los Angeles Times*, March 23, 1947, 11.
24. "County Given Research Fund," *Los Angeles Times*, March 26, 1947, A1.
25. This account of the tubercle bacillus and the clinical features of its infection of humans is derived from recent standard textbooks of medicine and of infectious diseases. One of the best is the chapter on "Tuberculosis" by Mario C. Raviglione and Richard J. O'Brien in *Harrison's Principles of Internal Medicine* (numerous editions). An excellent overview that mixes scientific, clinical, and historical approaches is William D. Johnston, "Tuberculosis," in Kenneth Kiple, ed., *The Cambridge World History of Disease* (Cambridge: Cambridge University Press, 1993), 1059–68.
26. Especially Jacob Riis, *How the Other Half Lives: Studies among the Tenements of New York* (New York: Charles Scribner's Sons, 1890).
27. See the discussion of nineteenth-century estimates of tuberculosis mortality in René and Jean Dubos, *The White Plague: Tuberculosis, Man, and Society*, with foreword by David Mechanic and introductory essay by Barbara Gutmann Rosenkrantz (New Brunswick, N.J.: Rutgers University Press, 1987), xiv–xv, 8–10.
28. On Keats, see Dubos and Dubos, *White Plague*, 11–17.
29. Frank Woodbury, "On the rational treatment of pulmonary consumption," *Philadelphia Medical Times*, July 15, 1882, 12: 693–703.
30. See Elizabeth Lomax, "Hereditary or acquired disease? Early nineteenth century debates on the cause of infantile scrofula and tuberculosis," *Journal of the History of Medicine and Allied Sciences*, 1977, 32: 356–74.
31. For example, James Frederick Brown, "The diagnosis of phthisis by the microscope," *British Medical Journal*, April 21, 1860, i, pp. 302–3.
32. Thomas M. Daniel, *Pioneers of Medicine and Their Impact on Tuberculosis* (Rochester, N.Y.: University of Rochester Press, 2000), 37–61.
33. Austin Flint, "Conservative medicine," *Buffalo Medical Journal*, August 1858, 14: 135–39; quotations from p. 138.
34. For an immediately pre-bacteriological view of tuberculosis etiology, and especially of the role of heredity and pre-disposition, see James C. Wilson, "The influences that predispose to pulmonary consumption," *Philadelphia Medical Times*, 1882, 12: 277–80.

35. Dubos and Dubos, *White Plague*, 33–43, discuss nineteenth-century examples of apparent heredity, including the Emerson and the Thoreau families.
36. John Blake, "On the climate of California in its relation to the treatment of pulmonary consumption," *Pacific Medical and Surgical Journal*, 1860, 3: 263, 351–472.
37. George Lewis, "The climate of the State of Minnesota, and its adaptation to persons suffering from phthisis pulmonalis," *American Medical Times*, March 15 and 22, 1862, 4: 147–49, 162–64.
38. Edwin S. Gaillard, "The climate and topography of Aiken, South Carolina, in their relation to phthisis," *Richmond Medical Journal*, 1866, 2: 16–23; Amory Coffin, "On climate in the treatment of pulmonary tuberculosis," *Boston Medical and Surgical Journal*, 1869, 81: 321–24.
39. John P. Wall, "The climate and diseases of the Gulf coast of the Florida peninsula, with remarks on the former in relation to pulmonary tuberculosis," *Charleston Medical Journal and Review*, 1874–75, n.s. 2: 107–26.
40. For medical theories justifying such locales, see Frank B. Rogers, "The rise and decline of the altitude therapy of tuberculosis," *Bulletin of the History of Medicine*, 1969, 43: 1–16.
41. Edmund Andrews, "The relations of cancer and consumption to climate in the United States," *Chicago Medical Examiner*, 1866, 7: 737–40.
42. Edmund Andrews, "Climatic relations of cancer and consumption," *Journal of the American Medical Association*, 1885, 5: 424–26.
43. Robert Koch, "Die Aetiologie der Tuberculose," *Berliner klinische Wochenschrift*, 1882, 19: 221–30.
44. Thomas D. Brock, *Robert Koch, a Life in Medicine and Bacteriology* (Madison, Wisc.: Science Tech Publishers, 1988), 27–139, esp. 117–39 on the discovery of the tubercle bacillus.
45. "The bacillus of tuberculosis," *Boston Medical and Surgical Journal*, April 20, 1882, 106: 377.
46. *Lancet*, April 22, 1882, i [119]: 655–56.
47. "Koch at the German Congress for Internal Medicine," *Medical Record*, June 3, 1882, 21: 603.
48. William F. Whitney, "The aetiology of tuberculosis," *Boston Medical and Surgical Journal*, May 25, 1882, 106: 487–91.
49. "Foreign Correspondence": W. T. Belfield, Vienna, May 2, 1882, in *Chicago Medical Journal and Examiner*, June 1882, 44: 613–17.
50. "Foreign Correspondence": W. T. Belfield, Vienna, July 3, 1882, in *Chicago Medical Journal and Examiner*, August 1882, 45: 170–73.
51. Henry F. Formad, "The bacillus tuberculosis and some anatomical points which suggest the refutation of its etiological relation with tuberculosis," *Philadelphia*

Medical Times, November 18, 1882, 13: 109–19; Henry D. Schmidt, "Microscopical investigation into the nature of the so-called bacillus tuberculosis," *Chicago Medical Journal and Examiner*, 1882, 45: 561–84.

52. Henry F. Formad, "The bacillus tuberculosis and the etiology of tuberculosis. Is consumption contagious?" *Maryland Medical Journal*, 1884, 10: 685–91, 705–11, 845–48, 865–69, 885–91, 905–11; Henry D. Schmidt, "The pseudo-bacillus tuberculosis," *Chicago Medical Journal and Examiner*, 1883, 46: 225–41.

53. Koch dropped a hint about his therapy into a lecture on "Bacteriological Research" given to the International Medical Congress, which was noted in detail by the correspondent of the *Boston Medical and Surgical Journal*, August 28, 1890, 123: 214. His full report, "Weitere Mittheilungen über ein Heilmittel gegen Tuberculose," *Deutsche medicinische Wochenschrift*, November 14, 1890, 16: 1029–32, was immediately translated in major newspapers, and published in the *British Medical Journal* and the *Medical News* of Philadelphia.

54. On Biggs' career and work see Daniel, *Pioneers*, 99–131, especially 113–19 on the contagiousness of tuberculosis and means to prevent it.

55. Ibid., 119–24.

56. For an overview, see Peter Warren, "The evolution of the sanatorium: the first half-century, 1854–1904," *Canadian Bulletin of Medical History*, 2006, 23: 457–76.

57. Edward Livingston Trudeau, *An Autobiography* (Garden City, N.J.: Lea & Febiger, 1916).

58. The female Drs. Gleason, who previously ran a water cure establishment in Elmira, New York, offered their patients a "sanitarium" of stone and tent buildings in the mountains at Las Casitas, four miles north of Pasadena. See "In the Sierra. Invalids Make Your Camps There and Win Health," *Los Angeles Times*, November 20, 1887, 2.

59. William M. Chamberlain, "Pasadena as a sanitarium," *Southern California Practitioner*, 1887, 2: 325–36.

60. The best popular exposition of the new roles recognized for vitamin D is Luz E. Tavera-Mendoza and John H. White, "Cell defenses and the sunshine vitamin," *Scientific American*, November 2007, 297: 62–65, 68–70, 72.

61. P. D. Davies, "A possible link between vitamin D deficiency and impaired host defence to *Mycobacterium tuberculosis*," *Tubercle*, 1985, 66: 301–6.

62. Two recent reviews provide an entrée into studies of vitamin D and tuberculosis infections: Michael Zasloff, "Fighting infections with vitamin D," *Nature Medicine*, April 2006, 12: 388–90; Kelechi E. Nnoaham and Aileen Clarke, "Low serum vitamin D levels and tuberculosis: a systematic review and meta-analysis," *International Journal of Epidemiology*, February 2008, 37: 113–19.

63. J. A. MacLaughlin, R. R. Anderson, and Michael F. Holick, "Spectral character of sunlight modulates photosynthesis of previtamin D3 and its photoisomers in human skin," *Science*, 1982, 216: 1001–3.
64. The effects of the zenith angle of the sun are discussed in Ann R. Webb, "Who, what, where and when—influences on cutaneous vitamin D synthesis," *Progress in Biophysics and Molecular Biology*, September 2006, 92: 17–25.
65. As an early example of multiple similar studies done around the world, see Ann R. Webb, L. Kline, and Michael F. Holick, "Influence of season and latitude on the cutaneous synthesis of vitamin D3: exposure to winter sunlight in Boston and Edmonton will not promote vitamin D3 synthesis in human skin," *Journal of Clinical Endocrinology and Metabolism*, 1988, 67: 373–78.
66. D. J. Turnbull, A. V. Parisi, and M. G. Kimlin, "Vitamin D effective ultraviolet wavelengths due to scattering in shade," *Journal of Steroid Biochemistry and Molecular Biology*, September 2005, 96: 431–36.

Preface

MISCONCEPTION has often been a potent factor in history. One particular legend, that southern California's climate offered a cure for tuberculosis and other serious diseases, influenced the development of the region certainly as much as have many solid facts. For a generation, tens of thousands of invalids, driven by the urge for self-preservation, flocked west and permanently changed the area. Historians have neglected this movement, partly because the numbers of these people and the dollars spent because of their coming, though large for semifrontier country, were few in comparison to California's other pioneer projects, such as gold mining, railroading, and petroleum development. And yet it is safe to say that the health quest was more significant for southern California's development than the great gold rush had been.

"Trying California" became more an act of faith than of desperation. Here was a story as dramatic as it was historically important. Although most health seekers came west with little money and no experience and followed new livelihoods under pioneer conditions, their contribution to the American westward movement was on the whole both courageous and successful. For those who blindly trusted climate, stark tragedy resulted in thousands of cases, but the majority of health seekers did survive, through following a more sensible way of life. Once they were well, they stayed on, built stable communities, and gave the area a far sounder basis for self-confidence than it had known. Victory born through misfortune is not unique in California's history. Men learn from failure, but most health seekers did better than merely "break even." They not only survived but materially helped make their chosen region one of the West's most advanced sections.

Being amateurs, the health seekers used terms loosely. For example, the words sanitarium, sanatorium, and health resort were employed interchangeably. In this book the author has applied the word sanitarium to refer generally to the health institutions. Sanatorium was often identified by health seekers as a retreat for tuberculars which stressed the curative powers of climate and a specific regiment, but the word has been used by the author only when it forms part of the name of a specific institution. The author has used the term health resort in a very general sense, just as the health seekers themselves applied it. To them, a community, a county, or all southern California was spoken of as a health resort. In this interpretation they were correct, for few health seekers in their generation were actually institutionalized. Consumption, a word infrequently heard today, was often used in the latter part of the nineteenth century. It specifically refers to tuberculosis of the lungs. In this volume, tuberculosis, the all-inclusive term, has been more frequently used.

Two aspects of the subject of this book have been treated in periodicals. The first, "The Health Seekers and Early Southern California Agriculture," was published in the *Pacific Historical Review*, XX (Nov. 1951); and the second, "Los Angeles County in the Health Rush, 1870–1900," was printed in the California Historical Society *Quarterly*, XXXI (March 1952). To these journals I express my gratitude for an initial presentation of the subject in a different form.

During my research on this topic, I visited numerous libraries and learned societies in California. Special thanks are due to the library of the University of California at Los Angeles, the library of the

University of Southern California, and to the Los Angeles County Medical Association. Mrs. Ana Begué de Packman of the Historical Society of Southern California supplied valuable reminiscences and let me use the society's newspaper files. I also wish to express my gratitude to the staffs of the California State Library at Sacramento and the Bancroft Library of the University of California at Berkeley for their help. Among the libraries of various communities, the writer especially appreciated being able to use the collections at Los Angeles, Pasadena, and San Diego. Mrs. Dorothy E. Martin and Mrs. Dorothy Watson on many occasions kindly offered their assistance at the Los Angeles County Museum.

I am particularly grateful to the trustees of the Huntington Library for making possible the publication of this book. Dr. John E. Pomfret's and the late Dr. Robert Glass Cleland's faith in the undertaking as well as their helpful criticisms are deeply appreciated. To Miss Mary Jane Bragg of Huntington Library Publications is due my sincere thanks for her careful editing of the manuscripts.

This volume could never have proceeded far beyond a mere outline for a book without the information supplied by several witnesses and participants: Dr. George H. Kress, Dr. Raymond G. Taylor, and Dr. Sumner J. Quint of Los Angeles, Dr. Francis M. Pottenger, Sr., of Monrovia, and Mrs. William Lauren Rhoades and Mr. C. W. Jones of Sierra Madre. Great thanks is due to the late Dr. Donald J. Frick of La Verne for his expert advice on medical matters.

For several years it has been an honor and a source of happiness for me to call John W. Caughey an inspiring teacher and a great friend. Dr. Caughey not only suggested this topic but offered invaluable advice on numerous occasions and examined the manuscript from first draft to finished product.

To my parents, Edward S. and Mary Louise Baur, I inadequately express gratitude for years of unlimited encouragement in this and every other endeavor I have attempted.

J. E. B.

Los Angeles, California
August 1959

The Health Seekers of Southern California

1870-1900

CHAPTER I

Building a Health Legend

To THE adventurous, California has always seemed a treasure house. Varying with time and the individuals involved, the coveted objects which California has offered have included strategic position, human souls to be saved, a great port, and rich veins of gold. Yet the most valuable of mankind's many physical treasures is health. The announcement that southern California's climate offered restored health to America's sick inevitably focused new attention on the area. For the first time world-wide fame came to the southern part of the state. The result of the advertising was a migration of health seekers, by which, perhaps for the first time in American history, a frontier was developed by the sickly and the invalid.

Unlike the more important discovery of gold, this knowledge was not suddenly learned one January morning. For three generations Spaniards and Mexicans had enjoyed the coastal climate of California and had been aware of the basic climatic differences between northern and southern California. In 1826 Governor José Echeandía temporarily removed the capital from Monterey to San Diego in order to improve his health in the milder southern region. Even before American acquisition, such visitors as the trail blazers Antoine Robidoux and Lansford W. Hastings were telling easterners about the wonderful climate.[1]

[1] Lansford W. Hastings concluded, "I cannot but think, that it is among the most favorable resorts in the known world, for invalids." *The Emigrants' Guide to Oregon and California* (Cincinnati, 1845; reprinted, Princeton, 1932), p. 85. J. Q. Thorn-

The glowing story was perpetuated by early settlers, although the pioneers who spread it were crude amateurs with scant background for interpreting what they saw into sound geographical and medical terms. Besides, like many highly impressionable people, they seldom agreed. For instance, in 1869 Charles Loring Brace believed that southern California's Anglo-Saxon population would eventually tend toward an Italian or Moorish type, due to the climate and a fruit diet. "A 'southern' aspect," he assured listeners, "is already very perceptible even in the pure Anglo-Saxons of Los Angeles and its neighborhood."[2] George Ward Burton had his own interpretation of climatic effects. He was convinced that California's atmosphere was modifying the vocal organs of natives and in time would make them "a race of singers" and southern California "the land of song for the Western Hemisphere."[3] Optimism was in the ascendant, but pessimists such as Hinton Rowan Helper dissented with a bitterness equaling the saccharine comments of the majority. In almost all instances climatic conditions were oversimplified.

The health legend might have been dissolved with amusing, exaggerated anecdotes preserved in western folklore, except that this story was not all fable; much of it was true, and, just as important, it soon brought an increasing number of health seekers to the golden land. A few of them were forty-niners. Several of the diaries and travel accounts of the gold rush documented California's growing legend of restored health. In 1876 Judge David Belden remembered the minor migration which was a part of the greater gold rush. He said:

> Who does not recall, in the adventures of the early days, the thousands that, smitten by disease, essayed the voyage round the Cape, the perils of the plains, with the cry "Health or a speedy grave." To not one in

ton and his wife set out for Oregon in 1846 "with the hope that its pure and invigorating climate, would restore this inestimable blessing we had long lost." *The California Tragedy*, ed. Joseph A. Sullivan (Oakland, 1945), p. 1. John Bidwell, leader of the first overland wagon train from the Middle West to California in 1841, upon his arrival in the West wrote, "Let me here remarked [sic] that those of the Company who came here for their health, were all successful." John Bidwell, *A Journey to California* (San Francisco, 1937), p. 44.

[2]*The New West; or, California in 1867-1868* (New York, 1869), p. 371.

[3]*Burton's Book on California and Its Sunlit Skies of Glory* (Los Angeles, 1909), I, 71, 138.

> a score of these came the coveted boon of health. Their resting-places mark the pathway of our empire from the banks of the Mississippi to the shores of the Pacific. . . . Suffering and disease were their companions as they journeyed hither. Pestilence and hardship welcomed their arrival. By the rivers whose golden sands had lured them from peaceful homes and loving friends, they fell by thousands, unknown and unremembered.[4]

Only a fraction of the gold seekers had been invalids, but while the sick were relatively few in ratio to the well, their real numbers were of some consequence. A group of them had fled the great Mississippi Valley cholera epidemic of 1848-49 only to find that it had arrived in the diggings before them. A minority of the rest were tuberculars, and as they came with few provisions, thinking California's climate uniform everywhere, scores or perhaps hundreds died in the chilly Bay region.

Before the promise of health could be made good, California would have to develop a new branch of knowledge, medical climatology. It would have to be well organized by trained scientists. Even before the gold rush had ended, considerable time and money had been expended in the study of California's mineralogy, geology, botany, and general topography through mining, railroad, and coast surveys, but a whole generation of American rule passed before much valuable knowledge had been gained in meteorology, a field indispensable to the effective development and use of health resorts. During the gold rush, J. Praslow, a German traveler, had made a beginning. He studied the temporary epidemics of the period, caused by the miners' improvidence and the sudden tremendous growth of towns without a corresponding advance-

[4] Quoted in George Wharton James, *Heroes of California . . . as Narrated by Themselves or Gleaned from Other Sources* (Boston, 1910), p. 488. William L. Manly, hero of the Death Valley party, who saw numerous health seekers on the trail in 1849 said, "many came not solely to obtain gold, but that which would prove of greater value, restored health." *The Pioneer* (San Jose), Feb. 15, 1894, p. 4. Another famous '49er, Alonzo Delano, was advised to try California. He relates, "besides the fever of the body, I was suddenly seized with the fever of mind for gold; and in hopes of receiving a speedy cure for the ills both of body and mind, I turned my attention 'westward ho!'" Alonzo Delano, *Life on the Plains and among the Diggings* (Auburn, N.Y., 1854), p. 14. For an account of a health seeker of 1854 who tried the overland trail to California, see Nancy A. Hunt, "By Ox-Team to California," *Overland Monthly*, 2nd Ser., LXVII (1916), 321-326.

ment in sanitation. Taking into account temperature ranges, the agricultural possibilities of California for dietary improvement, and the amount of sunshine, he predicted unexcelled health for a large population on the Pacific Coast. Unfortunately, Praslow's *State of California: A Medico-Geographical Account,* though originally published in German in 1857, was not translated into English until 1939.

While Praslow was at work, California established a Committee on Medical Topography within the state's medical society. That same year, 1857, Lorin Blodget's learned *Climatology of the United States* appeared. He cautioned that, due to the recency of California's occupation, it was "difficult to trace the climatological geography of disease there, but enough is known to decide that malarious diseases are comparatively rare, and that their antagonist forms as observed in the eastern United States, or the pulmonary class, are almost unknown from California southward." Even that early he had become confident that for recovery from respiratory diseases southern California's climate was superior to that of Italy. The mortality rate for tuberculosis reached its "absolute minimum for the continent in temperate latitudes . . . in southern California."[5]

On the West Coast, knowledge grew slowly. Finally, in 1869, J. D. B. Stillman, editor of the *California Medical Gazette,* called for a society to:

> centralize and organize effort and direct the line of observation, to preserve and collect results, whether little or great, and make them contribute to the desired end; data could be accumulated that in course of time would enable those who come after us to draw legitimate conclusions and not leave them at the mercy of that fabulous personage, the oldest inhabitant.[6]

Stillman had called for difficult work. Meteorology was still a new science and medical climatology little more than good guesswork. By then, however, the impetus of an increasing invalid migration had appeared, and in the 1870's there arose a group of volunteers eager to study the therapeutic values of climate and inform the health seekers on what had been found. To most of the writers who participated, this was

[5](Philadelphia, 1857), pp. 460, 477.

[6]*California Medical Gazette,* II (Oct. 1869), 34.

a crusade, for they themselves had come west as invalids and had found improvement or cure. Thus their approach was usually subjective. Yet, these amateurs were medical explorers. Their trails of discovery, the methods they employed, and the conclusions arrived at are more interesting than even the vivid men themselves. Their literary products, though both copious and sincere, oversimplified situations and sinned on the side of enthusiasm. Furthermore, the crudeness of that day's medical knowledge complicated their shortcomings. In southern California, most physicians were not acquainted with the latest discoveries regarding tuberculosis, the health seekers' chief disease. Fortunately, by the mid-eighties this lag would be corrected. Meanwhile, the medical world's achievements were almost immediately outmoded, due to the rapid broadening of knowledge about the disease between 1870 and 1900.

In consequence, the self-appointed advisers on tuberculosis exaggerated southern California's actually splendid climatic advantages. Many had been pioneers and, after a score of years in a particular region, felt they knew their locality to perfection. As a result, they sometimes sounded like quacks selling nostrums, the whole area being the panacea. In the long run, of course, their achievements in benefiting the ill were smaller than their hopes had been, though their labors stand out as a piece of effective publicity and an example of nineteenth-century optimism. With a greater body of scientific knowledge behind them, their more realistic successors of the next generation would do more for the tuberculars than these pioneers who had urged them to come westward.

These early writers, zealots and individualists all, nevertheless, had a certain uniformity of method. First, they tried to explain the geographical bases for southern California's advantages for health seekers, and in this endeavor they were reasonably successful and accurate. Dr. M. H. Biggs, a cautious man who weighed his words by the milligram, presented an excellent report to the state medical society in 1870 giving his findings, which would soon be copied by the boosters of climate. He explained that northern California lay more directly in the path of the winter cyclonic storms from the Pacific than did southern California, where the sudden east-west trend of the Coast Ranges south of Point

Conception protected the section from cold northwest winds. Coming south by sea from San Francisco, most visitors noticed the change of wind direction in the vicinity of Santa Barbara and the welcome warmth of that region. To the east the San Bernardino Mountains offered an added protection from desert winds. Biggs also claimed that the chain of islands along Santa Barbara's coast warded off cold ocean fogs.

The slant of the coast and southern California's mountain system had indeed tended to create a Mediterranean climate with relative aridity and the abundant sunshine basic to any direct relation of geography to the improvement of health. With good reason, sunshine was credited by all observers with many recoveries, primarily because it was "tempting to life in the open air." Inevitably, "sunshine chronicles" became something of a fad. For instance, L. Bradley, a former Illinois invalid, kept a chart of Santa Barbara's weather in which he recorded 310 days during which the afflicted could be outdoors five or six hours, 29 cloudy days, and 12 rainy ones, but only 15 days a year in which a health seeker would be confined to the limits of four walls because of severe winds or heavy rain.[7] Another volunteer precursor of the weather bureau prepared crude graphs from data gathered in all the important resort districts. Using hygrometer and thermometer, he discovered that Anaheim, where he had settled for health, put in the shadow both Menton, France, and Aiken, South Carolina, traditional habitats of the sickly.

Solar rays were not southern California's only attraction. There were cool nights, unfamiliar to most easterners who spent restless summers in sweltering heat. Refreshing sleep was a year-round blessing in this western area; it speeded recovery through proper rest. Yet, days were not always balmy. There were storms, too, and, surprisingly enough, many doctors argued that even this unevenness in the weather pattern proved a beneficial change! Dr. Peter C. Remondino compiled abundant data on humidity, impressive evidence that health and good climate are not dependent upon a minimum of moisture, provided the temperature is moderately low. A later generation of experts would agree with him. Wisely, he added that, despite the rainy season, southern California was

[7] *All about Santa Barbara, Cal.: The Sanitarium of the Pacific Coast* (Santa Barbara: Daily Advertizer, 1878), p. 92.

never humid. Present-day geographers term the area's climate dry subtropical.[8]

Nevertheless, there were variances from this opinion; one group of medical men insisted on the necessity of nearly uniform weather throughout the year while another believed that tuberculars needed a bracing climate with well-defined seasons to prevent enervation. Truth lay between them, as time and case histories would prove; these would also demonstrate that southern California was no hygienic utopia.

Case histories, too, provided early writers with material evidence concerning the illnesses the climate might help, although to what extent climate alone would provide a cure, they did not ascertain. After years of having treated the sick, physicians compared notes, read as widely as possible on the diseases most prevalent in their region, and yet often concluded with the most naïvely optimistic of pronouncements. Witness the following:

> Speaking broadly, persons suffering from any of the following conditions will find certain locations in Southern California to be useful aids in restoring them to health,—incipient or early phthisis or tuberculosis in any form, chronic pneumonia or a tardy convalescence from either pneumonia or pleurisy, diseases of the liver following malarial poison, cirrhosis of the liver, simple congestion or hepatic catarrh, jaundice, functional disturbances, and organic ills in those of advanced years and weak or poorly-nourished children, children subject to one of the various diatheses, as the strumous, rachitic, or tubercular. The overworked and overworried class will find here a most soothing climate to regain their lost energy or restore the nervous system to its normal equilibrium.[9]

[8]The Mediterranean climate of southern California is among the rarest of all major climate types. In summer the region is within the subtropical high-pressure belt, an area of dry, slowly descending air. Skies are clear. Breezes from the cool ocean current coming from the north keep the land cool for some distance inland. The land cools rapidly on summer nights through radiation. The pressure belts shift southward in winter, and southern California is influenced by the prevailing westerlies, bringing moderately cool air and occasional cyclonic storms. Thus the rainy season is November to March.

[9]William A. Edwards, M.D., and Beatrice Harraden, *Two Health-Seekers in Southern California* (Philadelphia, 1897), pp. 99-100.

At least the doctor admitted that he spoke broadly! His statement had its basis in truth, yet seems more a confession of faith than an analysis of proven fact. Sadly for posterity and tragically for their readers, many other writers wrote with even less restraint. Assuredly they often told the truth; a majority told nothing but the truth, as they saw it, but hardly any were able to tell the whole truth, then an unknown quantity.

One layman, stressing the climate's reputedly restful values, explained that "Nervous diseases here are shorn of half their terrors. All forms of mental disease find a specific in the soft but not enervating atmosphere that comes to the tired author like a sweet sleep."[10] Yet the atmosphere had done little to prevent the necessity of building a branch insane asylum in southern California.

Most writers on climate began their books and pamphlets with general estimates of southern California as a whole and further along dealt in detail with the attributes of their specific area, usually the author's home town. There were four important resort areas: Santa Barbara, chiefly the city, but to a lesser degree the whole county; Los Angeles and vicinity; the San Bernardino region; and the area immediately behind San Diego, port for several health resorts. These sections were comparatively thickly settled with invalids, and their fame became state-wide and even nation-wide through the writings of conscientious medical men as well as the less altruistic but sometimes more effective efforts of commercial health-resort promoters.

Almost every community had at least one imperturbable, and usually venerable, physician to champion its assets. Dr. Samuel B. Brinkerhoff, the "grand old man" of Santa Barbara medicine and chief booster of the local salubrity, was a classic example. Brinkerhoff had arrived in 1852 and for almost thirty years cared for the town's sick. According to one writer, he "knew as many family histories and family secrets of the section as a father confessor."[11] Brinkerhoff thought he knew almost everything about the climate, too. For years he had noticed near Santa Barbara a large body of petroleum rising to the surface of the sea and floating over

[10] [Jesse D. Mason], *History of Santa Barbara County* (Oakland, 1883), p. 456.

[11] Yda Addis Storke, *A Memorial and Biographical History of the Counties of Santa Barbara, San Luis Obispo and Ventura, California* (Chicago, 1891), p. 79.

it for several miles. Having read that in plague-ridden Damascus good results had recently been achieved by spreading crude oil in the gutters, he missed completely the elementary principle behind the incident. Actually, the oil had destroyed mosquito eggs, thus preventing fever. As Brinkerhoff saw it, the prevailing westerly sea breezes, passing over the expanse of ocean-borne petroleum, probably took up "some subtle power which serves as a disinfecting agent," and thus accounted for the infrequency of fever and the superior health of his fellow citizens.[12]

Confident of the logic of this respected physician, his prominent colleague, Dr. Thomas M. Logan, endorsed this explanation. Here is an extreme example, but it shows vividly the fundamental weakness in the methodology of so many of these men. Their honest conclusions were usually perfectly reasonable—on the surface—but they applied the deductive method in their interpretations of phenomena rather than a more scientific inductive process. Superficiality resulted.

The Los Angeles area produced the same type of amateur climatologist found in Santa Barbara. The former city's less picturesque setting probably accounted for the lack of so large a group of versatile defenders, but it was late in advertising itself as a health resort, even at the hands of tireless realtors. Citizens had early announced the salubriousness of their community's climate and for a time let the matter rest, for Los Angeles intended to become the region's metropolis, not dependent upon the invalid for development. Nevertheless, this type of settler came, though for most of the early period his praises were less quoted locally than those of his Santa Barbara counterpart. Los Angeles had little typhoid and malaria and no yellow fever. Catarrh, liver, and kidney diseases were uncommon, while neuralgia and rheumatism usually disappeared or greatly improved after one removed inland. Not only did the vicinity lack the occasional chills of coastal settlements, but it was safe for those whose heart condition might make living dangerous in the higher altitudes of most inland resorts.

[12]*First Biennial Report of the State Board of Health of California . . . 1870 and 1871* (Sacramento, 1871), p. 82. Subsequent numbers were published in *Appendix to the Journals of the Senate and Assembly of the... Session of the Legislature of the State of California* (Sacramento). Hereafter reports will be cited as CBH, and session and volume number of *Appendix* will be given.

In 1880 the state health board's "Committee on the Establishment of a State Hospital for Consumptives" was founded. The investigation it undertook gave worth-while publicity to the advantages of each California region, and especially to Los Angeles County. This was probably the first important study which was unbiased, comprehensive, and sound. The committee ascertained a given region's temperature, humidity or aridity, elevation, frequency of fog and wind, and water supply. The importance of most of these assets is self-evident. Water alone would make or doom several health-resort locations. Locally, the group solicited data from the most reliable medical men. Replies were kept brief and specific, avoiding as much as possible the almost traditional florid verbosity of that era. For a score of years these findings were quoted by those who would profit thereby. More significantly, this investigation stimulated as never before the study of California climate in its relationship to tuberculosis.

Although it had probably the best natural advantages for tuberculars, the sparsely settled San Bernardino territory did not receive so extensive study and publicity as the more thickly populated areas of southern California. Its share of attention was devoted to the hot springs.

San Diego was more fortunate in this respect. National figures as early as John C. Frémont and Stephen W. Kearny during the Mexican war had praised its sunshine. In gratitude to a benefactor and contributor to the health legend, San Diego might do well to erect a monument to Louis Jean Rudolph Agassiz. In 1872 the great scientist spent a short time there and before leaving said:

> There is one advantage that I, as a scientific man, may lay more stress upon than is necessary; but I hardly think it possible. It is the question of latitude. You are here upon the thirty-second parallel, beyond the reach of the severe winters of the higher latitudes. This is your capital, and it is worth millions to you.[13]

Shortly afterward, the area's chief apostle of climate, Dr. Peter C. Remondino, wrote *The Mediterranean Shores of America,* a contribu-

[13] Quoted in San Diego Chamber of Commerce, *Descriptive, Historical, Commercial, Agricultural, and Other Important Information Relative to the City of San Diego, California* (San Diego, 1874), p. 20.

tion to medical climatology, which revealed not only his findings but also his regional affections. Agassiz had told Remondino that San Diego was one of the earth's favored spots, and these few words gave encouragement to several local climatologists.

Preceding both Agassiz and Remondino, in 1870 Dr. D. B. Hoffman had prepared a study of the area's medical topography. More critical than most of his enthusiastic contemporaries, he admitted that fevers and rheumatism did exist, the latter attributed to coastal humidity. Meanwhile, Remondino was claiming that tuberculars who came to town voiceless and emaciated soon spoke normally, gained weight, and slept soundly. Apparently none improved enough to burst forth as opera stars, as had been predicted earlier, though Remondino insisted that "very often, with nothing but the hygiene of common sense and climate, the patient glides into recovered health almost insensibly."[14] Overcome by such fabulous claims, a visitor of the late sixties, William A. Bell, had been told and unquestioningly believed that death was "a remarkable event" in San Diego. The leading doctor, probably Remondino, had said that a physician who depended upon his practice would starve to death. With the diffusion of this claim, doctors would soon have plenty of patients!

In a general sense, San Diegans had reason to boast. C. M. Plumb, a layman, in 1873 prepared a list of ten criteria for judging a health resort. San Diego fulfilled his medical decalogue, for it had the needed dry atmosphere, equable temperatures throughout the year, salubrious sea breezes, cool nights, semitropical fruits for the invalids' diet, and "absolute freedom from miasmas."[15]

San Diego was also fortunate is having a United States Signal Service observer stationed there. The meteorological tables compiled at this post provided further data for the interpretation of medical researchers.

Almost invariably, the amateur climatologists of southern California included extensive tables in their works. Vital statistics were universally and convincingly employed, for low mortality rates are tremendous

[14]*The Mediterranean Shores of America: Southern California: Its Climatic, Physical, and Meteorological Conditions* (Philadelphia, 1892), p. 119.

[15]San Diego Chamber of Commerce, *Information Relative to the City of San Diego*, p. 21.

forces of persuasion. They alone must have brought hundreds of health seekers, in spite of the fact that they had usually been haphazardly gathered and their accuracy was highly questionable. Physicians set to work collecting information covering the period from 1782 to 1870 for births, deaths, and marriages in Santa Barbara. This careful research in mission archives and other repositories showed that by the latter year the death rate was 7.7 per 1,000 population, a gradual improvement through the generations, though Santa Barbara had always been a healthy place. Meanwhile, San Diego was claiming that in 1873 there were only 53 deaths for its population of 9,000, or 5.8 per 1,000, and 13 of these fatalities were of transient tuberculars.[16] At the same time New York had 32.64 and Philadelphia a rate of 26.28. By comparison, San Diego seemed almost antiseptically perfect. Taking advantage of such attractive exercises in arithmetic, it claimed to be America's healthiest "city."

Nevertheless, figures could not be trusted. Today's demographers assert that there were no reliable data on deaths in California until 1906. Furthermore, the death rate for tuberculosis became an increasingly weak point in southern California's arguments, though most fatal cases originated elsewhere. The sick might as well have stayed home; according to one pessimist, Santa Barbara and San Diego had as high a tuberculosis death rate as Westchester County, New York. Unfortunately, even accurate vital statistics cannot report recoveries. As one physician expressed it, the best doctors lose the most patients because they have the most, and this was true of "Dr. Southern California." Regardless of contemporary apologists, the figures remained high. Henry Harris, an author-physician of the twentieth century, found that during the seventies roughly one death in three was directly due to tuberculosis in San Diego and Santa Barbara and one in five in Los Angeles. The Bay area, which no longer attracted tuberculars by that time, still had one in every six deaths attributable to phthisis. Easterners doomed by the disease arrived by sea and rail first in San Francisco, and its cemeteries not infrequently claimed them before graveyards farther south could do so.

[16]*All about Santa Barbara*, pp. 8, 49.

To counteract bad publicity, the legend builders sometimes referred to spectacular cases of longevity. The examples of Indians who supposedly had survived to the age of 140 years were publicized and their photographs reproduced in guidebooks. Señora Eulalia Pérez, nearing her sesquicentennial, according to unauthenticated accounts, got more than her share of fame. Such cases seldom, if ever, can be substantiated; certainly they could not have been in the nineteenth century when birth records were scarce and those few in existence vague and faulty. Then in spite of the increase of numerous debunkers, as late as 1915 Dr. Marion Thrasher brought forth *Long Life in California,* the thesis of which was an argument for the positive effects of climate upon extreme longevity. When such a topic as this was discussed seriously by responsible medical men, their work took on an air of sensationalism and trespassed upon the huckster's fantastic realm.[17]

If the climate had been as near to perfection as these pioneer climatologists believed, they would have been recognized by posterity as significant contributors to medical knowledge. As it was, they did clear away much ignorance regarding California. They showed that the southern part of the state possessed not one but many climates and that there an individual could certainly escape from the extremes of weather found in most other parts of the United States. Persistently, they gathered data on health and climate and made way for the trained medical climatologists who would follow a well-defined and specialized subject with the better tools and methods of a later day. Trail blazers can seldom be expected to function as expert highway engineers, but they do serve the elementary need for courageous exploration and the presentation of broad, general concepts of knowledge. The doctors of the seventies and eighties deserve praise for their pioneering. They gave hope to America's desperate invalids and inspired many of them with enough of their own confidence to come west for new life in a fertile land. Unfortunately, the partly true legend that they had helped to create would be further distorted by other groups, quite often less sincere in purpose.

[17]As early as 1854 Pringle Shaw had noticed during a visit to San Diego "plenty of shrivelled up Indians to be seen tottering along, who have forgotten their age, but who were men and women ... more than seventy years ago." *Ramblings in California* (Toronto [1857?]), p. 102.

CHAPTER II

Advertising the Resorts

IT WOULD be both unjust and inaccurate to say that every man who pioneered in publicizing southern California's climate to health seekers had a bit of the brash booster in him, but the majority of those who reached a large audience certainly did. As we have already seen, their zealous efforts had uncovered useful facts, but with these had been accepted a large amount of misinformation based on inconclusive evidence and generalities. With the legend of an optimum climate already prepared for them by sincere and respected citizens, the commercial health-resort boosters now played their part. These men, whose motives were sometimes ulterior, had their ubiquitous characteristics. Shyness surely was not one of them. Often put forth in crude form, their literature almost always revealed a lack of medical knowledge. At best their endeavors could be of limited value, and sometimes their words were costly to health seekers in wasted time, train fare, and pathetic effort. Occasionally they proved dangerous, for while southern California was persistently peddled like a patent medicine, it was useless to sufferers from many diseases who might have been saved at home.

A specimen of this type of advertising was the pamphlet entitled *Southern California: The Italy of America*. Bearing neither date nor place of publication, it claimed for the area the "only perfect climate in the world and the grandest scenery under the sun" and continued:

> Invalids come here by the hundreds, and in every case, where they are not past all hope, they speedily find that precious boon which they have sought in vain in every other clime. . . . Consumptives, whom physicians of the East had declared past all help have come here and in a few weeks have shaken off the fetters of that Eastern ice-born curse.[1]

This pamphlet used the lethal psychology that the more glittering the generality, the better it might convince the gullible. Actually, at any given time or place, very few in the last stages did recover from tuberculosis. Such literature conveniently and cruelly evaded that basic fact.

Characteristic of the health-resort boosters was a spirit of intense rivalry, competition in the form of exaggerated localism. California's first important contest with Florida resulted. Almost every writer, whether he was a responsible physician or a questionable promoter, reviewed the qualifications of Europe's older health resorts and found them unsatisfactory. To these men, most of whom had never been abroad, the Old World fell short when its advantages were compared to those of southern California. In the same vein, other American retreats were branded inferior to those of the Pacific Coast. During the sixties and seventies South Carolina, Arkansas, Minnesota, Texas, Wisconsin, and Colorado had achieved considerable favorable publicity regarding their climate. These regions, however, were never so viciously attacked as was Florida, archvillain of the California boosters. Florida was undesirable, they said, because "The surplus of water, the lowness of the land, and the long, hot summer make Florida subject to malaria and fevers. . . . No part of the State is a favorable all-the-year climate for an Anglo-Saxon constitution."[2] But what a different situation one found in southern California! "In all my wanderings," a partisan enthused, "I have never before found a place where I would so gladly fix my permanent abode as in this Paradise in Southern California."[3] Dr. J. S. Adams, a physician, belittled Florida, which, he claimed, had "a humid atmos-

[1] Pp. 1, 8.

[2] R. W. C. Farnsworth, ed. *A Southern California Paradise* (Pasadena, 1883), p. 94.

[3] Banning *Herald*, Sept. 5, 1891.

phere, and more or less malaria."[4] Another insisted that anyone who settled there would suffer years of chills and fevers before he could become acclimated. The San Diego press played up the story of an invalid lady who wished "from the bottom of my heart it had been so that we could have gone to San Diego" instead of Florida, where, interestingly, she found frozen plants and shivering tourists.[5] For several years, Florida had been publicized by boosters, but despite its good head start over California, lodgings were poor and development long remained slow and uneven.

Among themselves, the boosters of southern California's pioneer health resorts disagreed almost as violently on the merits of their specific localities as they attacked Florida's claims. Once the potentialities of this invalid trade were sensed, region was baited against region. At an early date, detractors had predicted that Los Angeles would never be anything but a dusty pueblo because of its inadequate water supply, and they did not overlook the dangers of poor drainage in the hot months known as the "fever season." When the city had grown large, despite their prophecies, then Los Angeles was said to have too dense a population for the ideal accommodation of the sick.

On its part, Los Angeles answered back. Rather understandably, its rival for primacy, San Diego, was the chief object of Angeleno wrath. As late as 1889 this traditional jealousy was still rampant, and evidence of the ugly situation was easily found. During a two months' sojourn in Los Angeles, a National City visitor was unable to buy a San Diego paper. Meanwhile, an Ohioan was told by an Angeleno with whom he became acquainted that San Diego had "a constant and prevailing fog for the year round, causing malaria, diphtheria, and hundreds of other contagious diseases, and that he [a health seeker] had decided not to go but to locate in Los Angeles where it was more healthful."[6] The next year it was reported that the new Hotel del Coronado was a pesthouse, closed by quarantine officers about a hundred times, and pneumonia was

[4] William R. Bentley, *Bentley's Hand-Book of the Pacific Coast* (Oakland, 1884), p. 26.

[5] San Diego *Union*, Jan. 9, 1873.

[6] San Diego *Sun*, Nov. 18, 1889.

rumored striking down San Diegans at a terrible rate. Angrily, the *Southern California Informant* bragged that "San Diego is the most healthy city in the known world."[7]

The envy of Santa Barbara's neighbors had its origin in that town's rise as southern California's best-advertised health resort. The fortunate little city used as its weapon of counterattack a haughty condescension. William H. Seward, Princess Louise (the daughter of Queen Victoria), and a few years later Mrs. Benjamin Harrison were the community's guests; all praised the climate while promoters made the most of their statements. One local zealot remarked:

> Santa Barbara prides herself on being more æsthetic and cultured than her somewhat plebian sisters, San Diego and Los Angeles, and the impress of royalty that Princess Louise gave the city had a very expansive and exhilarating effect.[8]

Meanwhile, visitors to southern California expressed surprise, amusement, and disgust at these petty civil wars of words. One rationalized that, as inappreciable antagonisms are common to new lands, they would dissolve when population and trade had expanded. He forgot that once commerce has grown, great new rivalries may be stimulated. At that time the most important commodity on sale was a salubrious climate.

Almost unique, the Chino *Valley Champion* at the height of the boom attacked those "inconsistent journalists" who besmirched one part of California for the supposed benefit of another. The writer described the genus:

> He is a detestable creature that tries to raise himself by lying about others; likewise a journal is pursuing a contemptible course when it seeks to belittle or misrepresent a neighboring town or locality. Should the Eastern correspondents collect and print the unfriendly and untruthful assertions made by the journals of one part of our State about

[7] (San Diego), Aug. 24, 1889.

[8] Mary C. F. Hall-Wood, *Santa Barbara as It Is* (Santa Barbara, 1884), p. 19.

> other parts, they would do more injury than they can possibly do by any lies they can invent.[9]

Attack upon one's rival was but one side of the counterfeit coin, the tarnished side. The other side was polished brightly, by optimism, exaggeration, and boastfulness, which the boosters had concocted in publicizing a particular locality. Anything less than superlatives elegantly adorned would probably have been considered treason. In tune with his place and era, an overzealous Angeleno asserted:

> We have a tradition which points, indeed, to the vicinity of Los Angeles, the City of the Angels, as the site of the very Paradise, and the graves are actually shown of Adam and Eve, father and mother of men, and (through some error, doubtless, since it is disputed that he died) of the serpent also.[10]

Those of more serious nature were sometimes little closer to the truth. One boasted, "In all Southern California there is no spot so well situated sanitarily as Pasadena,"[11] while another exclaimed, "Pasadena is the greatest all-the-year-round health-resort in the world."[12] At least as early as 1866 a citizen of Los Angeles had adopted that title for his town.[13] The same honor was claimed by San Diego, San Bernardino, and Santa Barbara. To San Diego, the first part of its name might stand for "sanitarium," but neighboring communities spelled it San Di *EGOTIST*, or Sandy Ague. According to Fannie Gaylord, her town, too, had a destined duty:

> Fair Santa Barbara, to thee
> Is given a sacred ministry.

[9] Chino *Valley Champion*, Mar. 9, 1888, p. [4].

[10] *California as It Is*, 3rd ed. (San Francisco: San Francisco Call Company, 1882), pp. 166-167.

[11] Farnsworth, p. 98.

[12] Quoted in Katherine Abbott Sanborn, *A Truthful Woman in Southern California* (New York, 1893), p. 188.

[13] G. W. Barter said that Angelenos ought to erect signs "on all the roads leading into the city, warning physicians and undertakers of the danger of starvation attending a residence within." Los Angeles *Star*, Jan. 10, 1872.

To thee the sick and suffering
Their hopes and fears and sorrows bring.
Would those sad hearts so sorely tried,
Might see their longing satisfied![14]

Not to be bettered by the folk culture of a rival, Los Angeles had its own self-appointed bards; all were fervent, but all are forgotten. One of them, Thomas F. Joyce, had arrived in December 1882 because of a lung malady incurred in saving two girls from drowning in a Massachusetts millpond. In verse he eulogized the City of Angels:

The good Lord, He made this a dear land,
Where pleasure and health they clasp hands.
Here the sunshine of heaven
In abundance is given,
So that man can God's work understand.[15]

Other would-be poets penned similar hygienic stanzas. It is little wonder that a reaction began.

The ballyhoo of the resort boosters soon became detrimental to southern California's reputation as well as dangerous to trusting health seekers. Consequently, a reaction against this preposterous propaganda was begun by southern California's friends, those who knew the area well and realized that it had much to offer the newcomer. As early as 1877 Dr. F. W. Hatch attacked extreme optimists as well as the hucksters, complaining:

> A good deal has been said and written about the climate of California as a home for the consumptive, and the most extravagant opinions have been promulgated, particularly by non-professional travelers, as to the marvelous virtues of certain portions of the State.[16]

In 1881 Joseph J. Perkins, a reputable Santa Barbara promoter, said of the evils of extravagant boasting, "much . . . has been well calcu-

[14] *All about Santa Barbara,* p. 49.

[15] Los Angeles County Pioneer Society, *Historical Record and Souvenir* (Los Angeles, 1923), p. 134.

[16] Fourth CBH (1876/77), p. 38, in 22nd Sess., III.

lated to mislead those living at a distance. I do not mean to say that the writers themselves designed to mislead or *misrepresent,* but if such *had* been their design, they need not have chosen a different method."[17] Others believed that the many advantages of living in southern California were apt to make one oblivious to any evils. Angered because at least half the invalids in the last stages of tuberculosis were supposed to be perishing soon after their arrival, one observer blamed doctors as well as boosters and overenthusiastic tourists:

> A locality which has once attained the reputation of a health resort has to bear the blame when unfavorable results ensue, and the gross ignorance or brutal indifference as to the fate of the invalid, which was the primary cause of allowing the long and hopeless journey, is usually forgotten. With every allowance for whims and moods, no honest physician will fail to tell his patient, unless there are excellent reasons to hope for recovery, the best place is at home.[18]

Rumor, propaganda, and inadequate advice had created for the eastern health seeker a strange picture of southern California. Many of them expected to find a land free of rain, fog, wind, and clouds, endowed with 365 days a year of warm sunshine and enjoyable cool nights, offering a cure for all sufferers in the last stages of any malady imaginable and a remedy for every trouble in Pandora's box or Satan's mind. Disillusionment was inevitable, as the following illustrates:

> People have actually entered San Diego Bay in the morning, intending to spend the winter, and left for home the same evening without getting off the steamer, simply because it was raining. Others fly from Los Angeles to Santa Barbara, expecting to find it drier farther north, and, reaching there before the storm has finished, leave for some other place.[19]

Realists had to enlighten the naïve. Never really organized and without a definite program, beginning haphazardly and concluding at no

[17] Joseph J. Perkins, *A Business Man's Estimate of Santa Barbara County* (Santa Barbara, 1881), p. [3].

[18] George Rice, *Southern California Illustrated* (Los Angeles, 1883), p. 34.

[19] Theodore S. Van Dyke, *Southern California* (New York, 1886), p. 201.

fixed date, conscientious doctors, civic-minded residents, altruistic newspaper editors, and hundreds of "seasoned" health seekers took it upon themselves to inform inexperienced visitors honestly but often with harsh frankness. This chore was a quiet and apparently little-remembered aspect of the health-resort publicity campaign.

First of all, who should come to southern California? Most earnest writers agreed that only persons in the first stage of tuberculosis would derive perceptible benefit from the climate. Those in the second stage might possibly gain a few months of life and a measure of physical comfort, but victims in their final sufferings, with lungs almost destroyed, would merely endure the financial and physical expenses of a long trip with homesickness and a coffin at the end of the railroad. Candidly, they were told to "stay at home and die in comfort"—and this by a former health seeker, Theodore S. Van Dyke, who had found in California his only hope for survival.[20] Still, Van Dyke was more brutally accurate than the day's imperfect medical statistics.

As we know today, many other types of pulmonary affliction actually could be aided by rest, sunshine, and pure air. Almost any delicate person not in critical condition might find definite help in southern California, but as wise writers even in the eighties cautioned, he must consult a reliable physician in the East to determine what region was particularly suitable to his condition. Unfortunately, this counsel was not always valid, for many of the most reputable doctors on the eastern seaboard knew little about conditions in southern California. As a result, an editorial in the *Pacific Rural Press* advised health seekers themselves to make a careful study of climate. To help them to do so, it called for a "popular hand-book describing places, precautions, and results already obtained . . . furnishing all necessary information as to precise locality where each should go to receive the greatest benefit to any particular ailment."[21] The next decade brought several fairly adequate studies. Meanwhile, local medical advisers believed that as any decision would certainly result in a sacrifice to the individual, it would be wiser to leave

[20] Ibid., p. 202.

[21] Quoted in Pasadena *Union*, Sept. 8, 1887.

one's family, friends, and career during the first stages of tuberculosis in order to live extra years. Then, too, physicians had earlier led invalids to believe that their western stay need not necessarily be permanent. Experience proved this advice wrong, for relapses were common among those returning to harsh eastern weather; some sufferers even died on trips north to San Francisco. The exhaustion of moving about killed many a tubercular, yet the quest for health gave the state quite a fluid population. Adept at coining succinct truisms, Drs. Walter Lindley and J. P. Widney said, "It must not be a trip, but a migration."[22]

A National City physician offered his opinion on the best way this migration might come to the coast. To him, the Santa Fe Railway route with its great variability of altitudes was a "fever-chart" line. It did in fact resemble such a diagram, as much therapeutically as in profile, for many health seekers had been made desperately ill by the overland route. Having been on trains when invalids en route had died, he was sure that the mountain heights were to blame, since the victims had seemed to rally until elevation increased. The central, or Union Pacific, route was also partly mountainous. Therefore the only solution, and the one often followed in later days, was to make the trip in several slow stages. This was somewhat of a reversion to the old covered-wagon itinerary, a lengthy trip which had less often proved fatal. As a summary of all this good advice, the moral was come early, come slowly, and come to stay.[23] In New York, a tuberculosis specialist always warned against any transcontinental trip during the winter, advising the Panama or Cape Horn route for cold months.

The second important question was, once arrived, what should the invalid do? Charles Nordhoff, himself often accused of being champion of the boosters, but at least one with integrity even if that virtue was mixed with undue zeal, counseled the new arrival to find the "oldest

[22] Walter Lindley, M.D., and Joseph P. Widney, M.D., *California of the South*, 3rd ed. (New York, 1896), p. 65.

[23] One observant health seeker described a tubercular's trip as follows: "I never saw such devoted attention as she had. They deadened her sensibilities with opiates, then exhilarated her with whiskey and champagne, and finally stimulated her with effervescing compounds." Mary H. Wills, *A Winter in California* (Norristown, Pa., 1889), p. 14.

resident doctor" and inquire about the dress, food, and lodgings which would be of the most value to the migrant health seeker. Food and housing were real problems. Eulogists of local conditions had written reams about low prices and rents, the abundance of cheap, wholesome food, and the ease of acquiring a comfortable standard of living in southern California. These assertions had in general been incorrect when early resorts were boomed during the 1870's, and they were almost universally untrue by the nineties. "California of the South is not the country for a poor invalid," warned Dr. William A. Edwards in 1897:

> One must not come here seeking health without sufficient means for himself and his family, or care-takers. I have seen so much distress and suffering on this account that I wish to speak plainly. . . . Everything that the invalid requires is expensive here, much more so than in the far East. One *can* live very cheaply, but only by denying himself the comforts and luxuries which are essential to the well-being of an invalid.[24]

Fresh foodstuffs were not expensive, but coal, wood, gas, water, ice, and all manufactured articles cost on an average about 40 per cent more than in the East. Furthermore, until late in the period of the health quest, outside Los Angeles, San Diego, and Santa Barbara, southern California afforded few of the comforts found farther east. This was a frontier, and the invalid must live accordingly, fully forewarned that doctors and drugstores were not always available. The health seeker was in a real sense a western pioneer.

In the years before sanitariums became common, during the late 1880's, accommodations for invalids were often bad and seldom beneficial. The sick would have to avoid first-floor apartments, especially in winter. Dr. Remondino offered special advice for daily living, cautioning:

> *If you are after lost health, attend to that, and don't convert yourself into a tourist or a picnicker,* as you probably require more rest and quiet

[24]Edwards and Harraden, pp. 96-97. D. L. Phillips had earlier warned that only invalids of means should go to San Diego.

> than revelry or deviled ham-sandwiches, hard-boiled eggs, and picnic pies or cakes. Make it your sole business to attend to what you left your home and friends for—the regaining of your health. Keep out of crowded hotels, churches, and parties; eat regularly, go to bed regularly, dress conveniently, live generously, be patient; do not expect to be transformed into a Samson or a Goliath in three weeks; expect ups and downs; stay in the open air as much as possible; be rational.[25]

If such wise recommendations had been followed, southern California would have observed fewer funerals and fewer suicides by invalids. Significantly, tuberculosis experts six decades later would stress diet and rest to the virtual exclusion of climate as a cure.

Among the advisers were advocates of outdoor living for health seekers in less critical condition. Their enthusiasm for the curative effects of camp, stream, and trail life was apparently unlimited. The great majority of southern California's health seekers certainly never saw the inside of a sanitarium. Every important doctor recommended taking advantage of maximum sunshine, warmth, and fresh air, plus a wholesome diet and plenty of mild exercise. Critics denied many claims of the boosters, but they had to admit that the sunshine was genuine, and any place where continuous outdoor life was possible tended to help sufferers from a multitude of ills. Even before the era of psychoanalysis, boredom and homesickness were interpreted as contributing factors to invalidism and relapse. An active life was the answer, it was thought, for the fairly strong asthmatic, tubercular, and dyspeptic.

Though excessive exercise probably killed more than the praiseworthy climate saved, J. J. Perkins, a Bostonian who came west in 1875, attributed his recovery to:

> living almost altogether, and sleeping much of the time, in the open air; exploring the cañons, roaming over the foothills, through the valleys, climbing the mountains, fishing, shooting and, in a word, entering upon any out-door exercise that was calculated to withdraw my thoughts from my physical condition.[26]

Another such refugee left Massachusetts in 1889, weighing 107

[25] Remondino, p. 127.

[26] Perkins, p. 5.

pounds, and slowly recovered from tuberculosis in Ojai Valley. He lived chiefly on milk, a cow being his only permanent companion for many weeks.

Tenting was the logical solution for many invalids' problems. A large, warm, open grate or stove and heavy clothing would provide adequate protection under most circumstances. There were few worries for the gypsylike health seeker who chose to live in the wilderness and travel about the coast ranges or through hidden southern valleys. Some thought one could even do without a tent, especially if he were an adult; only children and advanced sufferers needed housing. Most camping parties were wise to bring along healthy persons to care for the sick in case of crises and to perform the heavier tasks. An even more agreeable means of travel was hit upon, the house-wagon, forerunner of the modern trailer. Southern California's trailer residents of today might envy these pioneers who journeyed far in comfort and saw a fresh and wonderful country while they recovered.[27]

At the newly established village of Pasadena in June 1874, one invalid whose account of his adventures was published solved his own housing problem by camping out. Throwing a carpet over the limb of a tree, he made a tent-home for his wife and daughter. After six weeks, when he had grown strong enough to build a real house, the family gave up with a sigh the surprisingly jolly life.

Isaac Mast, an itinerant minister, wrote his *The Gun, Rod and Saddle; or, Nine Months in California* "to encourage some desponding invalid not to give up the struggle" but to follow his method of recovery from typhoid pneumonia.[28] From December 1872 to October 1873 Mast hunted in the mountains, rode through the plains, and fished most of the streams of central and southern California. With like manner and aim, a more famous health seeker, Theodore S. Van Dyke, wrote *Flirtation Camp* (New York, 1881), the fictitious adventures of a young Bostonian consumptive who achieved victory over both ennui and tu-

[27] Dr. Robert Glass Cleland recalled that at the century's beginning transient health seekers traveled about southern California in these house-wagons. The desert was usually their destination. Interview at San Marino, Nov. 28, 1955.

[28] (Philadelphia, 1875), p. 3.

berculosis by way of the great outdoors. Interestingly, another Theodore, who became a president, was about to follow the same regimen in North Dakota.

While the southern California press published unnumbered accounts of recovery in the wilds, George Wharton James, one of California's greatest exponents of closeness to nature, advised a tubercular to retire to the desert free from luxury and stimulants, where air and sun baths might cure him. A mind occupied with nature's endless beauties, he philosophized, could never dwell on disease and death.

Unfortunately, not all health seekers could abandon civilization. Still, urban life afforded considerable exercise, too. In the nineties, a rising young journalist, Charles Dwight Willard, recommended bicycle riding for its therapeutic values. Like other semi-invalids, he joined the "smart set" in bicycling, a fad for many Angelenos but a prescription for the sick. As the century changed, "wheelmen" increased in Los Angeles, where more bicycles were sold in proportion to the population than in any other city in the world. Suburban roads were good by then and resorts relatively close by and numerous.

Despite the valid counsel of realists in California, easterners had become vocal and unfriendly critics of California's climates. Even Van Dyke, a booster, admitted that "All talk about ozone, electricity, and such things in the air seems equally foolish. There is, probably, no positive agent in the air here that is not equally effective in Boston or Milwaukee."[29] Dr. Dio Lewis, a prolific writer and a nationally famous physician, enthusiastically supported New England's claims to healthfulness and insisted that southern California offered nothing to a Yankee. He claimed that hundreds of California's invalids had "found in New England climate a vitalizing sanitarium"—the health rush in reverse. Furthermore, camping out in New England could be as well managed and would prove as successful as on the Pacific Coast. Having spent some time in the Sierra, he presumptuously assumed himself an authority on the subject. Lewis counteracted the apologists for southern California and attempted to explode the health legend which had been developing:

[29] Van Dyke, *Southern California,* p. 208.

> During the last thirty years, I have seen, perhaps, as much as any man in the country, of consumption and its treatment. I have known hundreds to leave New York and New England for Florida and Southern California. I have induced many hundreds to remain here, dress in flannels and live in the saddle.[30]

Lewis claimed that stay-at-home Yankees had more frequently recovered than had large numbers of their fellows who had fled to milder climates. In this assertion he was a precursor of present-day physicians who de-emphasize climate in the treatment of tuberculosis. At Santa Barbara Lewis had seen "a crowd of consumptives from all parts of the Eastern States sitting about in the shade of piazzas and trees," but local doctors told him that, despite the idyllic setting, few improved and fewer recovered their health. This was often true, but Lewis erred in his insistence that, "Give Massachusetts the climate of Southern California, and in one year she would begin to lose her most precious treasures—the force and enterprise of her people."[31] His conclusions ostensibly bore more weight than those of most belittlers of California whose western stay had been limited, for he was a national figure. Others less famous, however, came closer to the truth.

Of course, there were chronic grumblers, people who were never satisfied, such as the man who scoffed that the probability that one would not be benefited by the climate was greater than that one would, and the disappointed health seeker who declared Los Angeles to be "one of the unmasked humbugs, time being the only essential to its exposé" as California's unhealthiest spot.[32]

Thus, the reaction to the health-resort hucksters on one hand took the form of constructive criticism and helpful advice to immigrants and, on the other, denunciation of southern California and all its works.

Conscienceless propagandists, however, could not be so easily silenced. They were primarily promoters to whom the health quest was a business, and business was good. These publicity men worked for sever-

[30] Dio Lewis, M.D., *Gypsies; or, Why We Went Gypsying in the Sierras* (Boston, 1881), pp. 406-407.

[31] Ibid., pp. 408, 404.

[32] Los Angeles *Express*, May 27, 1875.

al types of vested interests, groups dependent upon the prosperity of their holdings in southern California. The latter group included numerous real-estate promoters, landowners, railroads, newspapers, hotels, and development companies.

As was typical of the press in a new country, newspapers were always ardent promoters of southern California. They not only printed occasional articles in the form of visitors' testimonials to a given town's attractiveness as a health resort but frequently published handbooks, such as the Santa Barbara *Advertizer's* optimistic *All about Santa Barbara, Cal.: The Sanitarium of the Pacific Coast,* printed in 1878; the San Diego Chamber of Commerce's *Information Relative to the City of San Diego,* printed by the San Diego *Union* in 1874; and Arthur Kearney's *San Bernardino County: Its Resources and Climate,* printed in 1874 by San Bernardino's *Guardian.* The Los Angeles *Star* began to boost its small city in the sixties and for the next few years published long editorials with an appeal to invalids.

In 1890 Lewis Munson, editor of the Banning *Herald,* got up a 24-page pamphlet filled with facts and statistics on his locality as a sanitarium. The Redlands *Citrograph* applauded his work as "One of the neatest, oddest and most charmingly written brochures that has come to hand lately" and proposed that Banning mail 50,000 of these "health tracts" so that easterners could appreciate the advantages of the vicinity.[33] Partly as a result of information diffused in this manner, for a generation Munson's town was almost solidly constituted of health seekers.

The newspapers aided the health-resort boom for several reasons. Civic interest was always a strong motive. Editors and publishers sincerely wanted to benefit their community and see it grow. Most of them were also investors. Several, like E. N. Wood of the Santa Barbara *Index,* had come west as invalids themselves, entered the field of journalism, and when recovered felt they had an obligation and a peculiar opportunity to inform the public. A few, such as Ben C. Truman, were especially interested in railroad advancement and land sales.

Whatever the individual purpose, almost every editor served effectively in still another way. As most newspapers received a seemingly

[33] *Citrograph* (Redlands), Jan. 25, 1890.

endless flow of mail from easterners seeking advice on where to settle for health, the editorials inspired by such letters were calculated to appeal to the sick in general. A typical "tenderfoot's" query was: What are good farming lands worth? Is it a good climate for curing pulmonary complaints?

Of equal if not greater influence was the eastern press. Sometimes it published material scissored from southern California papers but quite often brought to light letters and interviews of invalids still in or just returned from the health resorts. In 1887 a southern California journalist who knew the Atlantic seaboard well remarked that the magazines and newspapers of the East had published countless articles on California, "mainly by visitors who came to pass a winter on account of their health."[34] He also noted the large number of Los Angeles periodicals which had been mailed by satisfied health seekers to friends back home.

Of course, less idealistic newspapermen wrote copy expressly for their advertisers, extolling the climate but using a conspicuously large amount of space lauding hotels and restaurants which health seekers should patronize. When brochures were prepared, these commercial establishments were profusely and gaudily illustrated by large cuts.

Real-estate agents and others interested in land were exceedingly alert to every opportunity. For example, Benjamin D. Wilson, a prominent pioneer of Los Angeles County and an important contributor to the region's development, wanted to donate building sites for a sanitarium which would have helped increase the value of his other holdings. The proposed Oak Knoll sanitarium-hotel never materialized, but it was characteristic of a type of project which was often successful.

In the eighties and nineties the Southern Pacific Company led several real-estate promotion campaigns. In an effort to sell its government grants, this powerful institution produced pamphlets such as *California: Its Attractions for the Invalid, Tourist, Capitalist and Homeseeker*. The railroad, seeking to sell less desirable properties, catered to invalids. At Indio, a potential site for settlement by tuberculars, hotel accommodations were still poor in the nineties, but with foresight the railroad had put up several three-room cottages so that tuberculars would not come

[34] George W. Knox in the Los Angeles *Tribune*, May 5, 1887.

into direct contact with other invalids. This solved two problems, those of housing for invalids and convenient transportation to the resort.

The various railroad excursions developed during the seventies and eighties were both effective vehicles of publicity for southern California as a "sanitarium" and a means of cheap transportation. Hotels were connected with these well-planned tours, and the railroads co-operated financially and otherwise. The Raymond excursions, which helped bring new settlers, issued several attractive pamphlets telling of its twelve routes terminating at Pasadena. Excursion trains left Boston, a rich source of raw material for the invalid trade. In providing both propaganda and transportation, the Whitcomb tours were almost as active.

Always energetic in publicizing its part of the state, the Los Angeles Chamber of Commerce advertised the health of the region for some decades after its second founding in 1888. In 1891 it offered a prize for the best list of one hundred questions to be distributed to eastern readers. Of course, each interrogation was accompanied by a cheerful answer, and most of these touched directly upon southern California's climate and salubrity. Number eight on the winning list was a good example of these statements: "How are lung and throat troubles affected?" it queried, and replied, "Favorably, they are alleviated in the worst cases [!] and cured in many others." Number nine ran: "What is the effect upon incurables?" The answer: "To relieve pain and render invalidism far more endurable than under other climatic conditions"—a polite way of calling California an opiate.[35]

Even foreigners played a part in making America and western Europe aware of southern California. In 1890 a Briton spoke of the great interest in the region among his countrymen. "Talk California there any time and you have a listener," he said.[36] Visiting southern California during the boom of the eighties, Samuel Storey, a member of Parliament, intimate associate of Andrew Carnegie, and himself a newspaper publisher, wrote a series of letters to the British press touching on west-

[35]Banning *Herald,* Sept. 26, 1891. The Chamber's *Facts and Figures concerning Southern California and Los Angeles City and County* (Los Angeles, 1888) emphasized the claim that here was the "healthiest city in the world."

[36]San Diego *Sun,* Jan. 14, 1890.

ern health resorts and other local conditions. This typical example from his columns informed readers:

> This is a land of promise for those threatened with, or suffering from, consumption, asthma, throat diseases, dyspepsia, or physical prostration. . . . Infectious diseases are scarcely known. . . . I refrain from saying more, lest a shipload of invalids arrive ere I depart. Suffice it to add that when once full knowledge of this incomparable land has reached our European physicians, it is not to Algeria, Madeira or the Canaries that they will send their patients, as a last chance.[37]

In the manner of a thorough booster, he listed several "remarkable recoveries." Thereafter, the journals of southern California claimed Storey as their own, quoting him and reporting the influence of his words in the United Kingdom. In 1889 his greatest effort on behalf of the Pacific Coast was the optimistic work *To the Golden Land: Sketches of a Trip to Southern California,* published in London.

A few hundred Canadians were to be found in the influx of health seekers, especially after the mid-eighties. E. Dastrous, correspondent for several French-language papers in the United States and Canada, came to San Diego on his doctor's orders in 1890. While there he prepared a series of articles on the region. Dr. Remondino supplied him with much data. Meanwhile, Ludwig Salvator's laudation of Los Angeles, *Eine Blume aus dem goldenen Lande,* had appeared in 1878, capturing the attention of central Europeans.[38] In comparison to those from the British Isles and Canada, there were few continental Europeans among the foreigners who came to California primarily for health.

American visitors returning to their homes in the East were far more efficient as walking advertisements of the health resorts than any inanimate paper could ever become. Their person-to-person influence would convince many. To draw attention to the new Mount Lowe Railway, George Conant traveled through the Middle West and New England in 1893. He reported scores who wanted every scrap of information about Pasadena and the cures they had heard were achieved there. Conant believed that he had convinced hundreds that California had

[37] Fallbrook *Review,* Aug. 7, 1891.

[38] Salvator's work was translated by Marguerite Eyer Wilbur in 1929 as *Los Angeles in the Sunny Seventies,* and republished in that city.

tremendous possibilities and an excellent climate. He brought back with him on the train several of his converts. Meanwhile, a Chicagoan returning home after having achieved full recovery found that his only annoyance now was having to tell so often the story of health on the Pacific Coast.

Through these various forms of publicity, southern California became well known throughout the nation with surprising rapidity. Partly because of the incoming health seekers, population increased, total income rose, and the health legend took on the aspects of a minor epic.

Yet even among its friends southern California's climate inspired humor.[39] At an editors' convention in Pasadena in 1886, one journalist of 240 pounds demonstrated by a joke how widespread and effective was knowledge of southern California's prime boast. "My beloved hearers," he began with a broad smile and mocking conviction,

> I thought I came here to die. Alas! when I left home I had but one lung and it almost gone. I couldn't speak above a whisper, and had no appetite. I have been two weeks in Pasadena, have three lungs, can roar like a descending avalanche, ate three mules for breakfast, and am going to try it for another week.[40]

This caricature of the invalid's testimonial, heard for a generation, proved as nothing else could the success of local advertising.[41]

[39] Peter A. Wagg, pseud., *Southern California Exposed* (Los Angeles, 1915) is an example of latter-day booster material humorously calculated to "expose" southern California as guilty of minor flaws which add up to something just short of perfection.

[40] *Porcupine* (Los Angeles), Mar. 20, 1886. Los Angeles could laugh at itself. The odors of a cesspool on Wilmington Street caused the local press to parody the mineral water craze, alleging that a physician had discovered "miraculous certain properties in the deposits contained in this pool" and recommending that a sanitarium be built. The Oakland *News* took the story seriously and republished it under the heading "An Important Enterprise." Los Angeles *Daily News*, Oct. 18, 1871.

[41] Probably the most famous and perhaps oldest story derived from the health legend is the one Edwin Bryant heard at Independence, Missouri, in 1846 where wagon trains were assembling to depart for California. In the original version, the hero was a 250-year-old Californian, perfect in mind and body but bored with life. As suicide caused damnation, he sought another means to die. He was advised to leave California's perfect climate and move to an unhealthy foreign city. On doing so he soon sickened and died. In his will, however, he had provided that his body be returned to California. There, the "health-breathing zephyrs" restored life to the corpse, and he submitted to his fate and lived forever. Edwin Bryant, *What I Saw in California* (New York, 1848), pp. 16-17.

CHAPTER III

Los Angeles, Capital of the "Sanitarium Belt"

IN THE 1860's Los Angeles was still a frontier town with a notoriety for its outlaw element. More Mexican than Anglo-Saxon in culture, this "Queen of the Cow Counties" had been changed but little by the great gold rush. Now, settlers were beginning to arrive for whom Los Angeles and the county it headed promised peculiar benefits. These were the health seekers. Hope had brought them west. Yet, in their long-sought haven, sanitary conditions, basic for their recovery, were still abominable, for many residents used the irrigation ditches for washing and bathing. Dust, filth, and poor accommodations did not unduly deter the coming of invalids. One of the earliest to arrive was General James Shields, who, having just completed his term in the United States Senate, in midsummer 1860 arrived by overland stage. A wound received thirteen years before at Cerro Gordo during the Mexican war prompted Shields's coming. The long trip proved worth while, for he remained to become a prominent "old settler."

The new transcontinental stage which carried the general to Los Angeles brought but a trickle of health seekers. A substantially larger number came by sea. Although probably there were still only a few score of health seekers in town by 1869, by that year Angelenos were complaining of the influx of indigent invalids for whom their community could provide neither adequate housing nor a steady livelihood. When the celebrated Pacific railroad was completed in May of 1869, it began to turn the migration into a rush which would last more than thirty years.

In January of that year, an editorial in the Los Angeles *Daily News* entitled "Wonderful Growth and Improvement of Los Angeles County" declared that "The great desideratum, a genial and healthful climate, constitutes the chief attraction."[1] At the time, the boast was nearly accurate. Already unpleasant results could be cited; one physician said that he had never seen in any other state so much suffering from pulmonary diseases as he had found in southern California. "Oh! what a lot of coughing suffering mortals are coming here! Many too late. One man died in sight of the harbor," exclaimed D. M. Berry in Los Angeles during 1873.[2] During the 1860's Phineas Banning's crude port facilities at Wilmington had served hundreds of health seekers, and his stages provided their first transportation from harbor to the inland towns.

This increasingly large migration for health could not have occurred at any earlier date, and the reasons for its occurrence changed within a generation. Before this period, the American standard of living had not been high enough for such a mass movement of invalids across the continent. Los Angeles, until the eighties a village surrounded by dry mustard fields and based on a vineyard and fruit economy, was hardly the place most of these people would have chosen had they been well. By the 1880's, however, the town was rapidly becoming a city in miniature, and railroads were ending its isolation. The promise of health had already played a noticeable part in making these fundamental changes, for although the health seekers constituted a minority of the newcomers, that minority was an active one, and either directly or indirectly it influenced a growing number of well people to migrate to California, many of these latter individuals being members of the invalids' families. By the nineties the sick had become as typical as the palm or orange. A tubercular described the result in Los Angeles:

> At every street corner I met a poor fellow croaking like myself. I strolled into the Plaza, there to imbibe the exhilarating effects of a

[1] Los Angeles *Daily News*, Jan. 15, 1869.

[2] D. M. Berry to Helen Elliott, Los Angeles, Nov. 24, 1873, Berry papers, Huntington Library.

> community with broken lungs in all stages, and the inspiring comfort of a vocabulary like this: "Well, how do you feel today? Did you have a good night? Are you trying any new medicine?" "O, the pain in my side is very bad." "Do you cough much now?" Fancy how helpful to an invalid! Then, at the boarding house opposite ours, one would come on to the porch muffled in shawls, a parchment face and sunken eyes, cough, cough, cough, leaning over the hand-rail another fellow with spindle legs and shrunken form; another joins them and still another, feebly walking to and fro, alive to save funeral expenses.[3]

On the streets and in the trolleys or at the post office, where they invariably awaited cheering letters from home, one could meet these desperately ill yet courageous people with their "hollow eyes and still more hollow cough."[4]

From their first arrival, the invalids created a housing shortage in Los Angeles, one of the first situations of this kind the quiet little pueblo had experienced. Their very numbers stirred up considerable action, and the picturesque Mexican community began to disappear before the inrush. Beginning its campaign in 1874, the Los Angeles *Herald* determined to obtain new hotels to end the crisis. The "city," as even then residents chose to call Los Angeles, was in a "catyleptic sleep" and was complacently permitting San Diego and Santa Barbara to build hotels and capture its potentially profitable health seekers. A 300-guest hotel, the *Herald* said, could be built for $175,000.[5]

Yet, in 1882 one health seeker noted that there still was urgent need for a good family hotel. Invalids were leaving for other towns because of this lack. Not all of them, however, were frightened off by poor accommodations, for the average had a strong backbone to accompany his weak chest and was willing to endure numerous hardships if there was

[3] Ontario *Observer*, Jan. 11, 1889. See also W. Jarvis Barlow, "Report on Two Hundred Charity Cases of Pulmonary Tuberculosis, under Sanatorium Treatment at Los Angeles," *Transactions* of the American Climatological Association, XXIII (1907), 164-165. This work shows the high percentage of foreign-born among the health seekers after 1900. Barlow estimated that 40 per cent of those treated at his hospital were aliens.

[4] Los Angeles *Herald*, Jan. 9, 1878.

[5] Ibid., Aug. 18, 1883.

the slightest hope of recovery. That same season the sick had filled all the available hotels and rooming houses, and many private families were taking in boarders for the first time. Real-estate men were reporting that houses had been bargained for before the foundations had been laid, and other observers concluded that health seekers currently constituted the major portion of the American population. Whether this was a valid assumption or not, their numbers were certainly greater than had been expected. Chief headquarters for the health seekers of the early eighties were the city's two leading hotels, which, unfortunately, housed only 250 guests.

Until later in the eighties, amenities were rare. The lack of them was bad publicity for Los Angeles. In one of its numerous editorials for better hotels, the *Herald* cried, "Let in the Sunshine and Stir Up the Fire!" Too many badly lighted rooms were being designed, despite ample evidence that:

> A large portion of the patrons of our hotels and boarding-houses, especially during the Winter months, are invalids, and persons in feeble health—persons who wish, and are willing to pay for comfort. The constant complaint among them is that they are compelled either to remain shivering in their rooms or else crowd around the public fire in office or parlor.

Hotel builders should watch the sun to determine at what hour it shone in certain directions and then construct their rooms accordingly. Flues should be used instead of pipes. In conclusion, the newspaper advised, "publish these facts—'fire and sunshine in all the rooms of the house'—and every steamer will bring you guests; you cannot drive them away."[6]

Only slowly was the housing problem remedied. While hunting quarters for an invalid friend in 1882, Dr. Norman Bridge could not find a single house in all Los Angeles providing means for heating in every room, except by gas or kerosene stoves, the fumes of which stayed in the rooms. Most places had only a kitchen stove or a single fireplace, usually found in the parlor. "They look upon you with horror and consternation here when you ask for a fire," observed a health seeker in

[6] Ibid., Mar. 29, 1874.

1886. His request eventually materialized as a small fireplace "the size of your thum-nail."[7]

At the time, the diet offered an invalid in Los Angeles was far from ideal, the abundance of food notwithstanding. Restaurant meat, bread, and cream were generally inferior, certainly not to be recommended for a tubercular who needed the best of nutritious meals. In boarding house, hotel, or restaurant, the sick had to accept the menu offered to the general public, for there were no health-food restaurants until after the turn of the century. Although visitors might vary widely in their praises of and complaints against Los Angeles, they almost uniformly condemned the city's twin shortcomings, poor heating and unsatisfactory meals. Confronted with these problems as well as a climate which offered less than the perfection they sought, it is no wonder that some health seekers were not surprised to find on a tombstone in Fort Hill Cemetery the inscription "Translated to a More Genial Clime."[8]

Other campaigns for civic progress were stimulated by those who felt that the health seekers were not receiving the welcome and the care to which they were entitled. An improvement in public sanitation was logically the first consideration of many citizens. The migrant sick had found their town far from clean. Scanning the newspapers of that era, one might conclude that Angelenos suffered from the world's worst sewerage system and that the appalling conditions were never improved from 1860 to 1900. This was far from the truth, although neighboring towns which had become Los Angeles' rivals would have had it thus. In Los Angeles, the editors, merchants, and board of trade used the issue of sanitation for propaganda, playing it up in order to get maximum action for their civic reforms. Unfortunately, their literature got beyond the city limits, spreading misinformation that they, least of all, wanted broadcast. Henry T. Finck, a tourist who wrote a book on his southern California visits as well as two widely read articles in *The Nation,* said that if Los Angeles wanted to retain its supremacy as a health resort it would have to

[7] Charles Dwight Willard to his mother, Harriet Edgar Willard, Los Angeles, Oct. 31, 1886, Willard papers, Huntington Library.

[8] Lloyd Vernon Briggs, *California and the West, 1881 and Later* (Boston, 1931), p. 122.

build a sewer to the sea. Most of Finck's general comments were favorable to the area, but this single remark was best remembered, despite the fact that he was a New York music critic and not a sanitary engineer. The scholarly editor of the *Southern California Practitioner* took exception to Finck's statement. He felt that the visitor had been influenced by Angelenos guilty of "blackening our own character."[9] Actually, Los Angeles was no longer either cesspool or pesthole. Its sidewalks were good, its streets sprinkled in summertime, and the drinking water better than that found in nine tenths of other towns of the United States. At the moment of his writing, Angelenos were enjoying the purest of spring and river waters. If the findings of the tenth census of the United States, made in 1880, are as accurate as one would normally expect, then Los Angeles was indeed well supplied with sanitary facilities.

In spite of the advancement of medical knowledge and sanitary precautions, as late as the winter of 1899 a smallpox epidemic of mild degree took a few lives in Los Angeles and drove off a good portion of the health-seeking trade. Smallpox scares had checked the influx of invalids several times before but had never stopped it. The health seekers' faith that this was a healthy city was not mocked. If one excepts the tuberculous newcomers, mortality was low; if one does not, it was still not excessively high.

For economic as well as altruistic reasons, the civic-minded of Los Angeles constantly sought to make easier and safer the lot of the health seeker. "Having insisted and brought them here, it [Los Angeles] should lose no effort to give them every possible sanitary protection now they are here," they argued. Specifically, one reformer objected to the open cars found during the eighties and nineties on the local street railway. These primitive conditions offered a pneumonia hazard during the winter; "it is a matter fraught with importance to this city as a national sanitarium."[10]

Even the "city beautiful," which did not become a significant crusade

[9] H. Bert Ellis, "Blackening Our Own Character," *Southern California Practitioner,* IV (1889), 18-21.

[10] Los Angeles *Telegram,* Nov. 8, 1893.

until the early twentieth century, got an initial boost toward prominence from the health seeker. Writing in 1869, an editor said:

> Strangers, wooed to our city by the great salubrity of our climate, and the well-founded reports of our wondrous resources, look in vain for the public parks, gardens and promenades beautified and adorned by fountains, trees and flowers, that add so much to the health and beauty of other cities.[11]

Although little was done in the succeeding years to unite individual efforts which were being expended toward community beautification, even in 1883 the *Herald* could agree:

> Our city is so especially the resort of the invalid that there should be every consideration for their comfort and enjoyment carried out . . . evergreen trees should give shade and odors of balsam; flowers yield their perpetual perfume; walks should be cleanly and smooth, and the streets kept free from dust and mudholes.[12]

With the boom of the eighties some of these amenities were provided. Streets began to be paved, and street lights appeared. The invalids had helped in the victory, for the growing population possessed a high ratio of these people, and the land boom was in part their doing.

For the development of Los Angeles, the greatest direct result of the health quest was in the field of sanitarium building. In 1872 the Los Angeles *Star* was calling for a sanitarium and recommending that "associations be formed to combine and build up such an institution, and not only will it be an act of humanity to the world, but it would be largely to the advantage of the locality."[13] Advocates of the plan suggested the hills north of town as the site for a new subdivision especially for health seek-

[11] Los Angeles *Daily News*, Jan. 23, 1869.

[12] Los Angeles *Herald*, Aug. 1, 1883. A visitor had philosophized, "It is true, there is a great deal of dust in the streets; but that helps to suppress the flies." Stephen Powers, *Afoot and Alone: A Walk from Sea to Sea by the Southern Route* (Hartford, Connecticut, 1872), p. 272.

[13] Los Angeles *Star*, Dec. 1, 1872.

ers who thus would be freed from the crowded conditions in an increasingly urban environment.

Probably the most grandiose scheme of that day was conceived by a colorful Angeleno, Frederick M. Shaw. Professor Shaw, as he liked to be called, had been a forty-niner. He claimed to have built the first three-story structure in San Francisco. This alleged feat was symbolically a good start for a man who so longed to build empires. After numerous unrecorded but undoubtedly exceptional adventures in the Bay area, the unconventional Professor Shaw came to Los Angeles and by the early seventies was firmly convinced of the community's endless future as the world's greatest sanitarium land. With superlatives as his sustenance, Shaw went to extremes from which even the most zealous boosters shied. Unlike them, Shaw had no ulterior motive. In April 1873 he interested prominent citizens of Los Angeles in his project for a unique health resort. Capital stock would be placed at $20,000. Shaw's proposed site for a modern Eden was a wise one, a section of the Santa Anita Ranch. Tirelessly, he expended much effort testing waters, taking temperatures, measuring wind velocities, and calculating the elevation of his chosen resort area. At last, the Southern California Sanitary Hotel and Industrial College Association began to take form in his mind. Its plan sounded like a medical constitution for utopia:

> First—The promotion of health and the art and science of preserving health.
> Second—The encouragement and prosecution of agriculture and the mechanical arts.
> Third—As auxiliary to said objects the acquirement of real estate and the erection of suitable buildings; said buildings to be arranged on improved sanitary principles, on land belonging to the Association, to accommodate the thousands of persons who are flocking to Los Angeles for health, pleasure or permanent residence. The plans of management involve the construction of dwellings, manufactories, waterworks, and the carrying on of all the employments that occupy the most enlightened people, and the furnishing of perfectly wholesome subsistence to all and any persons that may require it, at a less cost than it is possible to obtain these accommodations elsewhere.

If it had been practicable, such an institution would have been a godsend to the poorer invalids. Shaw explained that it would "freely demonstrate to all who are interested in such matters of health, that the most stubborn cases will yield to skill, patience and humanity."[14]

For a time, prominent men paid attention to this kindly eccentric, but as Shaw's association erected no buildings, interest soon waned. Still, he persisted in his idea, now demanding a tract of 5,000 acres in the San Gabriel Valley and promising that in three years a colony of health seekers would produce a $300,000 profit for an investment of $50,000. Basically, this plan was not irrational. While Shaw was dreaming, the Indiana Colony was being founded for like purposes, with a similar organization, and not far removed from his chosen site. Pasadena was the successful result of the Hoosiers' effort.

No great sanitariums appeared in Los Angeles during the seventies. A score of years earlier, however, the Sisters of Charity had conducted a hospital which succored the invalid regardless of his ailment. About 1886 this same Catholic group opened St. Vincent's Sanitarium in the hills overlooking the city. The sisters welcomed the sick irrespective of creed, and their rates remained reasonable. Meanwhile, the Los Angeles Sanitarium had been opened on South Hill Street. Advertising in a way calculated to appeal to the fastidious, it promised "a comfortable and pleasant Sanitary Home."[15] By the nineties Dr. Charles H. Whitman specialized in the modern tuberculin treatment at his Koch Medical Institute. With the increase in both numbers and standards of the medical practitioners in southern California, the lag in knowledge which had professionally separated the area from the progress made in the East and Europe had been considerably diminished.

The colorful boom of the eighties further emphasized the trend toward supplying the latest treatments for invalids and commercializing the resorts. That great phenomenon, as much psychological as economic, had as one of its few solid bases the promise of medical cure. It became

[14] Los Angeles *Herald*, Sept. 12, 1874.

[15] G. W. James, *B. R. Baumgardt & Co's Tourists' Guide Book to South California* (Los Angeles, 1895), p. [445].

standard practice for the realtors of the 1880's to erect flimsy buildings supposed to be used for sanitariums to benefit the residents of the proposed subdivisions. Some of them were on desert land. A local historian remarked that "When the fever of speculation was at its height it mattered little where the town was located. A tastefully lithographed map with a health-giving sanatorium in one corner, tourist hotel in the other, palms lining the streets, and orange trees in the distance . . . and the town was successfully founded."[16] Of course, the boosters' plans for sanitariums were merely daydream and diagram, but those who did come for health often stayed and gave greater impetus to the future growth, based not on wild speculation but on genuine resources of a potentially wealthy land. Theodore S. Van Dyke said of these people:

> Along with these came invalids and other climate-seekers, and people whose relatives here had been advising them to come out, and farmers by the hundred, tired of vibrating for seven months in the year between the fireplace and the wood-pile, dodging cyclones and taking quinine.[17]

Lured west by the boom before it had reached its zenith, a West Virginia evangelist told of the crowds of health and climate seekers who had swarmed into town. The era had all the noise, rush, and gaiety of a carnival, with some of the humor and pathos of a sideshow.

As early as 1883 the Los Angeles *Express* had to deny that its city lived solely on "climate and strangers," as many visitors thought. The newspaper insisted that residents of ten to twenty years had done the best and the most for the community, but even in this stable group there were a good many health seekers. The editor was basically right. There are no figures to substantiate anyone's claim, but perhaps 10 per cent of the people who came during the great boom had moved west in pursuit of health. Others came for speculation, for fertile farm lands, for "a change," or to retire. Another group, undoubtedly large, accompanied sick relatives. Yet, the health seekers' economic influence both before,

[16] James M. Guinn, *A History of California* (Los Angeles, 1907), I, 282.

[17] Van Dyke, *Millionaires of a Day: An Inside History of the Great Southern California "Boom"* (New York, 1890), p. 41. The best study of the boom is Glenn S. Dumke, *The Boom of the Eighties in Southern California* (San Marino, 1944).

during, and after the boom was certainly out of proportion to their numbers, as the newspapers' emphasis upon health seeking as a factor in land sales indicates.

Inevitably, faith in the omnipotence of a healthful climate added to the boomtime mania. One of a growing number of local analysts explained with some exaggeration:

> Los Angeles, Santa Barbara, San Diego counties have been practically developed and made what they are by the Eastern people who came out here for their health. They may be a "one-lunged crowd," as the facetious Missourians and old-timers dub them, but they have shown an amount of business sagacity and enterprise which puts the Californian to shame.[18]

Throughout the greater part of the United States the winter of 1885-86 was an extremely severe one. As a result of the cruel storms of that season the greatest of all migrations up to that time followed. Invalids entered Los Angeles no longer in small, isolated groups or as individuals but as a large band, a class with means. They mingled with the masses of rich tourists and greedy speculators and built fine homes in the uplands which had just been developed or on new tracts southwest of Los Angeles. A San Diego editor observed that their business acumen was sharpened by experience "to the keenness of a Sheffield razor . . . and what cannot a community become with classes like those to build it up?"[19] Another journalist agreed, marveling at their "handsome homes" which beautified the country. "We doubt," he summarized, "whether in the whole United States, so many intelligent and wealthy people could be found in the same area as found in Southern California."[20] Several years later David Starr Jordan echoed these booster statements, writing of scenes in his own day: "It is true that the 'one-lunged people' form a considerable part of the population of Southern California. It is also true that no part of our Union has a more enlightened or more enterprising

[18] Pomona *Progress*, June 3, 1886.

[19] San Diego *Sun*, Sept. 26, 1887.

[20] *Porcupine*, June 4, 1887.

population, and that many of these men and women are now as robust and vigorous as one could desire."[21]

Not the high intrinsic value of talented human beings but the extent and exercise of their money interested the boosters. A politician wrote of this veritable health rush:

> it differs from any immigration movement hitherto known to the West. The waves of immigration have been chiefly of young men who were home builders, poor in purse, and starting life on its threshold. This movement is of the middle-aged, of entire families, and transfer of homes already builded, of people of wealth, those who have reached the meridian of life and now want to lengthen their days and enjoy their too-often impaired health under the most pleasant conditions possible. . . . It is not a movement of money-seekers, but is one gigantic effort of prisoners to escape their life environments.[22]

Similarly impressed by the uniqueness of the movement, the prolific author Charles Dudley Warner maintained that:

> The movement of people thither is, both in quality and volume, the most striking phenomenon of modern times, in its character a migration perhaps unprecedented in history. It quite equals the movement of 1849, perhaps surpasses it in speculative excitement, but its original motion is entirely different. There was mixed, in the hegira of 1849 to the west coast, a greed for sudden wealth and a spirit of reckless adventure, which recalled the romantic heroism of both Jason and Cortez. The present emigration is not for adventure at all, and primarily not for gold; it is a pursuit of climate.[23]

A Pasadenan remarked, "These people know that, had they remained in the East, they would have been a subject for a funeral director"; Los

[21] David Starr Jordan, *California and the Californians* (San Francisco, 1907), p. 44.

[22] Los Angeles *Tribune*, May 5, 1887, reprinted from the San Francisco *Examiner*.

[23] Charles Dudley Warner, *On Horseback: A Tour in Virginia, North Carolina, and Tennessee, with Notes of Travel in Mexico and California* (Boston, 1889), p. 317.

Angeles should be grateful for their coming, for "this will be the world's greatest sanitarium for years to come."[24]

An editor in Riverside realized that southern California's land prices were almost unique, since worth was not always determined by the productivity of the soil but rather by climate. He demonstrated:

> Given, a man in Springfield, Ohio, who has consumed one lung in making one million of dollars; add a place in Southern California where the climate is so kindly that one lung is as good as two in Ohio, and he will to his living lung sacrifice a share of the million which the dead lung earned. All that part of California is the lung of the stricken East. Men go not to buy land but to buy lungs. . . . The boom may burst, but the climate won't. Nature runs a bank that never breaks, and she is in business down there, hence the boom. It may check; it may slip back, but it will gather force again. If you do not believe it, look at the jealousy of Florida, and listen to the gnashing of teeth along the Riviera.[25]

The doubting editor of the Los Angeles *Commercial Advertiser* did not share this optimistic conviction. Having little faith in the by-products of the quest for health, he believed that Los Angeles possessed only "adobe deserts" for a hinterland. Then, too, there was little industry in the region:

> Yet real estate is furious in its activity, even the people, who, in answer to questions, admit that there is no natural occasion for anything more than a small town in this quarter, are nevertheless convinced—or profess to be so—that the present purely fictitious prices of land will be maintained or increased. They found their faith on pulmonary consumption. Their argument is that the east is full of people nursing tubercles, who will continue to pour into South California in a steady stream to keep the boom going. For my part I don't believe it.

In the vicinity of Los Angeles there were lots enough to supply 100,000 eastern invalids with homesites, and people with property

[24] Pasadena *Daily Union*, Nov. 3, 1887.

[25] Pomona *Progress*, Nov. 3, 1887, from an editorial in the Riverside *Press*, Nov. 1.

were suspiciously willing to sell it, the only buyers being tuberculars, of which the writer frankly said, "I cannot believe in consumptives as a safe crop. There are fashions in climates for consumptives, for one thing, and Southern California may go out of fashion, as Minnesota did a generation ago."[26]

Meanwhile, Warner chuckled that "the buyer, amid the myriad signs of 'Real Estate For Sale,' ought not to be confronted by so many legends of 'Undertakers and Embalmers.' It chills ardor." Sarcastically he observed that southern California was trying to illustrate the converse of the theory of the survival of the fittest. The critic could be a good prophet, too. Like many others, he asked:

> What will all those people now there, and on the way there, do when they have sold out all the land to each other, and resold and resold it at constantly mounting prices, until it is beyond purchase, and it is found that no possible crop on it can pay a remunerative per cent. on the irrigated principle? What interests the philosopher is the inquiry, What sort of a community will ultimately result from this union of the Invalid and the Speculator?[27]

The collapse Warner had foretold occurred in 1888. Although the speculators may have fled, most health seekers, who had not been foremost in producing the illogical boom, remained in the region. While tens of thousands of speculators and climate seekers left Los Angeles, the center, brain, and purse of the land boom, the health migration continued into the next decade. Although Angelenos had been sobered by the collapse in paper values, few would as yet share the pessimists' view of the invalid and his larger destiny. Dr. A. E. Winship probably typified citizens of all Los Angeles County when he surveyed the aftermath of speculation in 1889:

> Pasadena has had her boom and it has burst, but all that ever could have made her great, grand and beautiful abide as equable, as genuine, as reliable as ever. It was a sad day when these acres that should have

[26] Los Angeles *Commercial Advertiser*, Dec. 8, 1886.
[27] Warner, pp. 316-317, 319.

> been sacredly set apart as the sanitarium of the world, that always welcome the weak, worn, and weary of earth, were thrown into the turmoil of financial strife, to wreck fortunes and crush hearts.[28]

Evidence that many enthusiasts still had faith in a predetermined Bethesda, observations similar to the following were made in the nineties:

> If all were to dry up, leaving nothing but our climate, this region would not be entirely uninhabited. People would come here and camp on barren rocks for the benefits to be derived from our health-giving climate. . . . Last year [1893] tourists brought more money to California than the entire fruit crop amounted to, and this portion of the State got the most of it.[29]

Even a decade after the boom had ended, health seekers partially accounted for the tremendous and largely unexpected new growth and prosperity. Sackett Cornell, canny publisher of the Los Angeles *Evening Telegram,* cautiously admitted that "the universal growth of the city in the past eight years is very largely to be ascribed to this one reason."[30] These newcomers were mostly permanent residents, although the city had by then a floating winter population estimated at at least 20,000, a large proportion invalid, which greatly augmented the winter income of Los Angeles.

After the boom subsided, invalids began to be exploited by boosters less universal in their appeals. Specialists in tuberculosis had arisen by that time, and their advertisements in southern California's newspapers systematically attempted to win the mass of health seekers. The boom had left much booster spirit that could not be easily liquidated. Thus, there was a widespread expansion of health resorts, some of high quality but a number of them unfit for their tasks. During the nineties the scores of haphazardly planned "health homes" and "institutes" began to decline. For the first time there were large establishments requiring extensive planning and big investments. The best of these had well-trained

[28] Pasadena *Daily Star,* Oct. 31, 1889.

[29] Needles *Eye,* May 12, 1894.

[30] Los Angeles *Evening Telegram,* Nov. 8, 1893.

staffs and building programs which alone would require careful preparation. For the founding of health resorts the tendency was to look to organizations rather than to individuals. Los Angeles, the real metropolis of southern California, logically became the nucleus of all this activity.

Each proponent had his own concept of an ideal resort, which varied from the small cottage project to a "hygienic state," a sanitarium-colony fully provided with all the political, economic, and social trappings of a modern city. To one reformer, his spacious establishment would cater to vegetarians and fruitarians or to those who preferred fasting. Others advocated electric light, sun, mud, massage, and sweat baths. A Los Angeles hygiene instructor recommended a sanitarium specializing in "scientific vital breathing." On the whole, the trend was toward individual treatment by methods fitted to one's peculiar ailment and its history.

In 1902 the great Barlow Sanatorium was incorporated as a nonsectarian organization to care for the county's indigent tuberculars. Dr. W. Jarvis Barlow, a prominent physician and through his hospital one of southern California's most effective philanthropists, gave Los Angeles its greatest sanitarium. The institution grew rapidly, its twenty-five acres adjoining Elysian Park in the Chavez Ravine offering space for many patients. Tent cottages and the provision of mild outdoor labor made the hospital partly self-supporting. Like Shaw's long-forgotten scheme, it took full advantage of air and sunshine.

In Los Angeles, the back-to-nature movement began to receive the abundant encouragement which it still enjoys. Many who had come in delicate health turned to the Battle Creek Institute's methods. Dr. J. H. Kellogg had established a movement connected with Seventh-Day Adventism in religion and related to vegetarianism in medicine. Before 1910 his institute in Michigan had become exceedingly popular, and its crusade for diet reform had been carried throughout the United States and into foreign countries. In southern California one could find an ideal spot to place Kellogg's doctrines in practice. Here an outdoor life might be pursued most of the year, while the area provided a constant supply of both health seekers and the fruits and vegetables recommended for their recovery. The Battle Creek Sanitarium Company sent scouts to Los Angeles and in 1905 purchased an estate near Glendale which was remodeled into a modern sanitarium. It soon prospered. Independent institutions at Altadena and Mentone had already adopted the system in modified form.

The rise of the health-food business was stimulated by both the back-to-nature movement and by the health seekers in southern California. Native herbs of the Los Angeles area made this unusual commerce possible. For example, yerba santa, a medicinal plant, was found widespread in local canyons. Persons afflicted with catarrh and asthma, most of them in the region for the climate, were advised to gather the herb and steep its leaves like tea or chew them. In the late eighties the California Positive and Negative Electric Cough and Consumption Cure, produced in the area from local roots, was widely advertised, although, we may safely assume, without beneficial results to tuberculars. By 1900 the rapidly growing health-food business had turned from medicinal compounds to cereals. A visitor commented, "The amount of medicine sold on the Pacific Slope is significant of either stupendous credulity or stupendous ill-health. . . . And the children get more than their share of the drugs. The weakening of a general belief in the Great Physician has quickened faith in quacks."[31] Unfortunately, some firms were unethical, and in 1904 the health officer found several companies adulterating their cereals with sawdust. Meanwhile, Los Angeles' first vegetarian restaurant was opened.

Economically, the health-food craze was a minor factor in Los Angeles. The migration of invalids, however, had great weight sociologically. For several years at least the influx helped keep white-collar labor cheap. Other considerations must be noted. Los Angeles had its anti-union leaders such as Harrison Gray Otis, who championed the "open shop." The fact that the city was not so industrialized as San Francisco partly helps account for this. Besides, the clerical and various semiprofessional and professional groups to this day are not intensively organized

[31]Horace A. Vachell, *Life and Sport on the Pacific Slope* (London, 1900), pp. 65-66.

anywhere in comparison to the industrial workers, and this situation was certainly even more typical of that era. It would be an exaggeration to say that the health seekers perpetuated the open shop into the twentieth century, but it is obvious that they did produce semiskilled and skilled labor in surplus. Naturally, employers preferred healthy men who could work long hours efficiently. The tuberculars and other semi-invalids sought white-collar jobs not only because most of them were middle-class, semiprofessional men but because for them great physical exertion was simply impossible. Suitable positions were almost immediately filled. Even as early as 1874 a Tennessean who was visiting California warned his eastern friends:

> I cannot advise any indiscriminate move of all classes [of health seekers]. The towns and country are overstocked with lawyers, doctors, merchants and clerks, so there is not much room for these. The country needs bone and muscle, good farmers, carpenters, plasterers, brickmasons, blacksmiths and day laborers on the farm, all of which command good wages. Farm hands get $25 to $30 per month and board by the year.[32]

Invalids had already created this imbalance. A quarter century later it could still be reported that "the number of tuberculous invalids or recovered consumptives seeking employment is at times simply appalling.[33]

Even those of higher training suffered. Two historians wrote in 1897, "The professions are greatly overcrowded, more so . . . than is true of other parts of the world, due to the fact that professional men who break down from overwork are continually coming to this country as health-seekers, and after regaining their health never go home."[34] Even the Los Angeles *Times*'s eternally optimistic Midwinter Number cautioned the invalids and healthy alike not to expect "soft" jobs and high pay, for:

> On the contrary, this is probably the least promising city of the size in

[32] Los Angeles *Herald,* Nov. 25, 1874.

[33] S. A. Knopf, M.D., "The California Quarantine against Consumptives," *Forum,* XXVIII (1900), 617.

[34] Edwards and Harraden, p. 112.

> the United States for persons who are seeking light employment, in the shape of clerks, or bookkeepers, or anything of that kind, as well as lawyers, and doctors, and parsons, and other professional men, or for people who desire to run a small store of some kind. The reason for this is that Southern California is the Mecca for thousands of invalids, who are glad to make enough to pay their board and lodging while they recover their health. On the other hand, there is an active demand for mechanics of all kinds.[35]

As recently as 1913 this situation still existed. Overstocked with professional men and top-heavy with white-collar workers, the area offered sickly easterners most unexpected and disenchanting employment. One might "get a job as street-car conductor—lots of college graduates get the air that way—or he might drive a laundry-wagon if he wasn't too proud, or take care of people's gardens." The reporter of the preceding said that his neighbor had "a weak-chested Methodist ex-preacher to mow his lawn and trim his vines."[36]

An observer in 1899 noticed the large number of small businesses and "an extraordinary variety of occupations, due to the fact that much of the population is attracted here by the healthfulness of the climate, and each tries to follow the business he is accustomed to."[37]

Death must have been a common topic of daily conversation in the Los Angeles of the late Victorian era. As health seekers raised the mortality rates, cemeteries grew and multiplied, and undertakers were occupied. Yet, the health seekers, especially the doomed ones, developed an admirable spirit of camaraderie. For the majority there never was an end to hope, but even among the resigned many maintained an almost cheerful fatalism. As a newcomer, Francis P. Rowland observed of his own kind:

> It is curious with what indifference the majority of sick people one meets speak of their ills. They speak freely of tuberculosis, and hemor-

[35] Los Angeles *Times*, Annual Midwinter Number, Jan. 1, 1904, p. 26.

[36] Charles Francis Saunders, *Under the Sky in California* (New York, 1913), pp. 249-250.

[37] Wyllys S. Abbott, "Los Angeles and the Teachers," *Overland Monthly*, 2nd Ser., XXXIV (1899), p. 81.

> rhages, consolidations and cavities; they don't hesitate to inform you, that they are consumptive, and that they wish to live through a winter in order to give the climate a chance to do them some good.[38]

They could laugh at the repeated yarn about one less optimistic of their brethren who dejectedly asked a Pullman passenger en route to Los Angeles of what use he could possibly be there and was thereupon comforted with the intelligence that the natives were already planning to open a new cemetery with him.

The gold rush had produced many descriptions of the wanderer who, having lost health, wealth, and illusions, died unknown and was simply buried in the diggings. The health quest of one generation later produced startlingly parallel accounts, such as the following: "Most of the graves in which sleep the once lonely and needy, will be found marked with but a narrow board, and upon it inscribed the occupants' name, age, and the date of his death."[39] Quite often even such unpretentious preparations as these were not made! A modern writer, Carey McWilliams, has said that the health seekers made Los Angeles the mortician's paradise. His statement is as factual as it is dramatic. In Los Angeles was built the first crematory west of the Mississippi. A British tourist, Horace A. Vachell, recorded the typical attitude toward funerals in 1900:

> In Southern California, funerals are, like the Irish wake, a source of entertainment to the many who attend them. If the deceased happens to have been in his lifetime a member of any order . . . his funeral becomes a public function, a parade. You march to the burial-ground clad in the uniform of your order; a band furnishes appropriate music. . . .[40]

Children usually attended what the author called "these send-offs." Nevertheless, some writers have said that funerals were so common at the time that few but the exceptionally curious noticed them.

[38] Los Angeles *Times*, Jan. 1, 1892. Los Angeles was a healthy city; in 1891 the health officer reported mortality at 13 deaths per 1,000 inhabitants. Phthisis pulmonalis caused 174 out of a total of 835 deaths.

[39] Emma H. Adams, *To and Fro in Southern California* (Cincinnati, 1887), p. 77.

[40] Vachell, p. 62.

Architecture was somewhat modified by the health quest. Combined with southern California climate which inevitably led to building on the principle of outdoor living, the health factor encouraged outdoor sleeping, large windows, glass porches, and airy parlors, in some cases adapted from the health resort by private builders. In time, even adequate heating facilities were provided!

Whether the health migration helped or hindered Los Angeles the more requires a qualified answer. Not all elements can be placed on either scale. The rich were usually a windfall, but the invalid poor were costly to both county and city governments through the medical care required for them. Late in this period a theory was advanced that health resorts were definitely valuable instruments of social progress. Advocates of the proposition studied several foreign institutions in urban and suburban regions. The general conclusion as reported by the American Climatological Association was that sanitariums bring far-reaching benefits to a town. Population increases, public health improves generally, and property assessment as a whole rises. Business, too, is stimulated. This conclusion, of course, referred to health resorts in the narrow sense of sanitariums where patients are more or less secluded, not resorts in the sense of masses of invalids following an illogical, ill-directed existence in a growing city, as was the interpretation in early southern California.

No such study was made in Los Angeles when the migration for health was still an important factor in the city's development. Even had it been made, with statistics still faulty in the region, it would have been to a large extent inconclusive in any sociological sense. Yet, Los Angeles was changed by the presence of the health seekers, and those changes are still apparent today in a great city which long ago ceased to be an important health resort.

CHAPTER IV

Town Building

AS EARLY as the 1870's contemporary visitors began to point out that Los Angeles, like all growing towns, was becoming less and less ideal for health seekers. Outlying areas were more desirable and yet benefited from the city's commerce and comforts without suffering from its crowded conditions. The well-known California author, John S. Hittell, predicted:

> Within fifteen or twenty years we shall have a score of towns like those of the French and Italian coasts of the Mediterranean built for and filled with invalids from distant lands. There is an urgent need that we should soon have several such towns, accessible by rail, so that invalids shall have no excuse for not coming to our State. The time has not yet come to determine where most of these towns will arise. Among the sites most worthy of attention are the San Diego Mountains, near Julian; the San Bernardino table lands, west of town; the mountains in the northern portions of Los Angeles, Ventura and Santa Barbara Counties. . . .[1]

In time, all these areas became health resorts, but even before the journalist had made his prophecy, a future city had been created. Pasadena was the important community based upon this foundation.

The "Crown of the Valley" was born of "the longing of a frail Indiana woman for a less rigorous climate" and her dreams of health in southern California. This lady was the wife of T. B. Elliott, an Indianapolis

[1]Los Angeles *Express*, Feb. 14, 1879.

army surgeon and brother-in-law of D. M. Berry. Elliott was the leader of a little migrating party which settled the new town in 1874.[2] Of these founding fathers, a survivor of the pioneers later wrote, "Nearly all of the men had followed professional or clerical vocations in the east. Most of them had come to sunny California to benefit the health of some member of the family."[3] The exceptionally severe winter of 1873 brought widespread lung diseases to central Indiana, thus inducing a number of sufferers to determine to move to a better climate. After consulting newspapers from Florida, Texas, and California and scanning several guidebooks and pamphlets, they paid tribute to its booster literature by deciding on southern California.

Even the exact location of Pasadena was determined by the health factor. In 1873 D. M. Berry had been sent by the colony planners to find a well-watered and timbered tract for the proposed settlement. Exhausted more by long-standing asthma than from his lengthy trip, Berry spent a night at the Fair Oaks Ranch of Judge Benjamin S. Eaton, an estate just northeast of modern Pasadena. For Brigham Young revelation had decided the place for his valley city, but for Pasadena's founders asthma chose the spot! Next morning Berry arose, struck his chest, and exclaimed, "Do you know, sir, that last night is the first night in three years that I have remained in bed all night?"[4] He usually had to sit up in a chair to breathe. Hurriedly, the land scout wrote Elliott of the wonderful location, suggesting, "Send along Geo. Merritt [an Indianapolis promoter] or some other enlightened man to build a Sanitarium next the mountains. It would be filled in a day and every visitor would be an advertiser of our fruit to all sections of the country."[5]

During the next decade Pasadena grew slowly, its economy based primarily on citrus culture. The majority of newcomers were still semi-

[2]Harris Newmark, *Sixty Years in Southern California, 1853-1913* (New York, 1926), p. 448.

[3]Jennie Hollingsworth Giddings, *I Can Remember Early Pasadena* (Los Angeles, 1949), p. 16.

[4]Quoted in Lon F. Chapin, *Thirty Years in Pasadena* ([Los Angeles], 1929), I, 96.

[5]D. M. Berry to Dr. Thomas B. Elliott and J. M. Mathews, Los Angeles, Nov. 14, 1873, Berry papers.

invalid. E. D. Holton of Milwaukee found several of his sick neighbors in or near Pasadena settled on large acreages which provided both life and livelihood. Some professional music teachers and fine arts instructors had moved to the Crown City for their health and soon created the first cultural opportunities, to a large degree lacking in near-by communities.

Pasadena began as a temperance town. Certainly it was well behaved, for not a single imprisonment was made during the seventies. A lady health seeker had to smuggle in a bottle of wine, thereby breaking a local ordinance in order to obey her physician. The town council, however, stretched a point and allowed wine to be sold on medical prescription.

The Indiana Colony was never a dour sanitarium, though, for toward the gray-blue mountains one could often hear the notes of a huntsman's horn and the baying of hounds echoing as they coursed along near the arroyo. Christmas, the heart of the tourist season, was an especially colorful holiday, gaily celebrated by the settlers.

During the eighties the original mood of a transplanted Indiana began to disappear. There were still a good many Hoosiers, but they had fallen in ratio to other middle westerners, a large portion of these from Iowa. As Iowa was a wealthy farm state, its citizens could afford to travel, particularly in winter. Of those who came for the season, a large proportion stayed on, especially if they were in delicate health. "The region of Los Angeles," an observer could say in 1883, "is getting to be a young Iowa."[6]

A few years later another traveler, the Reverend Wesley Hagadorn, commented:

> We found many invalids here, both among the old, the young and the middle-aged. The middle aged, perhaps, predominated. The prevailing cause of invalidism among them all was throat and lung difficulty. We found, in fact, that the majority of the inhabitants were here primarily because of poor health, either on their own part or of some member of the family. Many have been in part or wholly restored, some have died, and others are gradually declining.[7]

[6] Pasadena *Chronicle*, Nov. 15, 1883.

[7] Wesley Hagadorn, "A Pasadena Letter," *Southern California Christian Advocate*, Aug. 15, 1887, p. 5.

Though these last words sound ominous enough, Susie C. Clark, a Bostonian tourist, could note that "No letter from Pasadena ever omits to extol this locality as a health resort."[8] Already there was a local saying that widows owned Pasadena, since their invalid husbands had died there. Until the eighties, most health seekers lived in their own homes or in boarding houses and hotels, but the population was growing too fast to be housed in such manner. Private "sanitariums" were mushrooming on the outskirts; they were haphazard in plan and in treatments offered and sometimes little more than tents or wooden shacks raised by self-styled doctors.

In the early days Pasadena's main thoroughfare, Colorado Street, was usually referred to as "Doctors' Row." The medical men along that avenue were foremost in organizing a hospital for the community, though their efforts only tardily achieved results. Not until January 1895 was a receiving hospital opened.

Altadena, too, produced some enthusiasts for sanitariums at a rather late date. About 1900 Dr. W. J. Geierman, who believed that nature was the best physician, personally blasted rocks, built roads, and laid out his ideal sanitarium. Growth was nevertheless very slow, but in 1938, more than a generation after he had begun, there were ten small houses and a central dining and assembly bungalow to serve asthmatic patients. Tuberculars were cared for at Altadena's La Vina Sanatorium, built largely through cash donations of civic-minded residents. Dr. Henry B. Stehman had conceived of the institution about 1905 when he saw the need for a tuberculosis hospital in a neighborhood already long filled with victims of the disease. Endowed with equal heart, ten philanthropists subscribed $3,000 apiece to pay for the land and buildings. In a decade Stehman had a village of rude but efficient structures and was attending sixty patients, mostly charity cases.

About the same time the Esperanza Sanatorium was being opened at the foot of Mount Lowe, overlooking Pasadena. There, the hospital was equipped with X-ray apparatus and other types of the latest scientific

[8] Susie C. Clark, *The Round Trip from the Hub to the Golden Gate* (Boston, 1890), p. 35.

appliances. Despite modern facilities, Esperanza was plagued with the old-fashioned handicap of other health resorts—patients insisted on overeating and would not surrender their breakfast eggs.

Near-by was southern California's most famous and probably most beautiful health retreat, Sierra Madre, a foothill settlement founded by Nathaniel C. Carter. In 1872 this energetic health seeker had settled near San Gabriel. Believing as thoroughly in the power of advertising as did Charles Nordhoff, he inaugurated a series of railroad tours in 1874. These were the famous Whitcomb excursions, between New England and Los Angeles. In 1881 Carter purchased from E. J. "Lucky" Baldwin 1,100 acres of the wilder northern portion of his Santa Anita Ranch. Adopting the name of the range for his town, in 1882 Carter subdivided the rancho and sold tracts at Sierra Madre to health seekers. "Carterhia," his beautiful new residence, was the center of activity for its owner's advertising campaign. Trains had been Carter's chief medium of success, but the lack of a branch line to his colony retarded its growth. Not until New Year's Day, 1906, did the Pacific Electric trolley line begin serving this town. Nevertheless, even in the early eighties the foothills surrounding Sierra Madre were inhabited by growing numbers of invalids.

The Sierra Madre Villa had been established in 1875 by William C. Cogswell, an artist, who beautified the 500-acre estate with orchards, vineyards, and well-designed gardens. Later, in 1880, the Hatch Committee, appointed by the state health board to discover the ideal health resort of California, declared that Sierra Madre and vicinity was the perfect site for a state sanitarium. The magic of this official publicity made the town and the villa almost immediately famous.

After a visit to southern California's best-known resort, the editor of the Los Angeles *Herald* thought that it looked more like a village than a villa and would be a worthy rival to Santa Barbara's vaunted Arlington Hotel. He added, "We deem it to be our duty to inform persons who are sojourning in Southern California for their health that their 'Eureka' is in the Sierra Madre Villa."[9]

[9] Los Angeles *Herald*, Aug. 28, 1877.

Together, the hotel and cottages contained fifty-four rooms, including music and billiard halls, and a glassed-in veranda over 200 feet long with sliding windows to provide sunshine for the invalid. In 1883 General William T. Sherman found it "the most attractive spot for having a quiet, good time, on the American continent."[10]

Through the pages of the Sierra Madre *Vista* one may find numerous but seldom monotonous squibs announcing the arrival of new health seekers who, finding the hotel crowded, filled and developed the adjoining town. Great Britain, France, and particularly Canada supplied a number of these people. As they had elsewhere, physicians and their families made up a large minority of the community's population. Even Los Angeles doctors contributed to Sierra Madre's growth by recommending that their patients move to the foothills. By then several businessmen were living there for recuperation and each day traveled by tallyho or coach to the city. As could be expected, Sierra Madre and Pasadena were rivals. One feudist claimed that Pasadena captured none of Sierra Madre's tourists; rather, Pasadena's health seekers could not be kept away once they saw the more favored town. He praised Sierra Madre's lofty panoramic views but condescendingly asserted that Pasadena was the poorest of these.

By the nineties a majority of the permanent residents of the town of 2,000 was composed of health seekers, many of them suffering from pulmonary diseases. After 1900 tuberculars continued to come, congregating in tent villages northwest of town. There, a permanent resort was attempted by those eager to profit thereby, but public clamor caused its removal. The unsanitary condition of the flimsy canvas cottages had become so great a menace by 1907 that it was decided to incorporate Sierra Madre so that restrictive measures could be applied.[11]

[10] Quoted in Newton H. Chittenden, *Health Seekers', Tourists' and Sportsmen's Guide to ... Health and Pleasure Resorts of the Pacific Coast*, 2nd ed. (San Francisco, 1884), pp. 38-39. By the nineties the Villa was no longer primarily a health resort.

[11] C. W. Jones, resident of Sierra Madre since 1905 and mayor in 1908 when these ordinances were passed, knew nearly everyone in town and said that barely a family was there for anything but health. Interview with C. W. Jones, Sierra Madre, June 12, 1951.

To the east, Monrovia had become another gathering place for tuberculars, primarily because of the persuasions of doctors who had moved there themselves. Dr. Francis M. Pottenger came in 1895 for his wife's health. Despite her early death, he loved the new land and in 1903 opened the Pottenger Sanatorium, a private residential hospital for tuberculars of all stages. It was a success from the beginning and the first modern sanitarium in southern California. At that time there were only sixty such hospitals in the United States; Dr. Edward L. Trudeau had established the first in the Adirondacks in 1884. Although removed from Los Angeles, Pottenger's institution attracted considerable attention because of its advanced methods and the high percentage of recoveries achieved there. The average number of patients was 100; the all-time high, 151 resident patients, was reached in 1920.[12]

It would be tiresome and somewhat futile to list the scores of small sanitariums found throughout Los Angeles County and adjoining regions during the seventies, eighties, and nineties. The great majority of them were little more than boarding houses and in no way comparable to Pottenger's establishment. They ranged in interest all the way from a colony of asthmatics housed at Newhall in the San Fernando Valley to sanitariums for tuberculars in the Pomona region. Anaheim was the headquarters in the eighties for the Societas Fraterna, a health colony organized by an Englishman, George P. Hinde. Hinde and his associates laid down a rather dreary doctrine for their new way of life. Their dogma forbade the eating of meat, eggs, dairy products, and bread; only raw fruits and vegetables might be consumed by the faithful.

Southernmost of California's health-resort regions was San Diego. At the opening of the American period in California history, San Diego was a mere village some three miles northwest of its present center, a sunny adobe town with a potentially great port but little good drinking water for its scarcely 700 inhabitants. In 1852 Dr. A. W. Winder, a military surgeon, met four tuberculous army officers there. Even then visitors and residents alike had faith that the local climate would cure almost any ail-

[12] Interview with Dr. Francis M. Pottenger, Los Angeles, June 29, 1951. He has since published *The Fight against Tuberculosis: An Autobiography* (New York, 1952).

ment, and in this particular case it seemed to fulfill all that was expected of it, for the soldiers all recovered and later won military fame.

In the 1850's San Diego was a regular stopping place of the Panama steamers to and from the bustling San Francisco area. Sailing ships bringing Argonauts and their successors around the Horn often anchored there. A resident of San Diego reminisced that "scarcely a person either entered or left California without obtaining a taste of the delicious climate of San Diego. And not a few—from that mere taste—were induced at a later day to choose this region as a health retreat for themselves or friends."[13] By the eighties the railroad had put southern California in swifter communication with the East. Gradually the number of winter visitors increased. Invalids and businessmen in need of a rest had heard of the sunny port and came to renew their vigor. Most of them were men of capital.

Douglas Gunn, long-time editor of the San Diego *Union,* helped create this favorable situation. His newspaper had begun in its first volume in 1868 to advertise the healthful climate. Gunn calculated that 100,000 invalids in the United States had enough money to travel extensively. About $5,000,000 of tourists' cash flowed into Florida annually. If San Diego were to be as extensively publicized as the former state, health seekers would bring their yearly income west, thereby contributing to the prosperity of local merchants and mechanics and yet seldom competing commercially with older residents. Other sick persons would settle permanently once it was universally known that malaria sufferers, asthmatics, and rheumatics could be helped, and San Diego would not require railroads to make it great.

Could a city be built on the promise of health alone? Gunn optimistically believed that it could and referred to San Diego's "fan mail" inquiring about the climate for permanent residence. "In none of these letters do the writers even speak of our prospects; health seems to be the only object they seek," he observed.[14] California's oldest town had the climate and the potential port. A railroad would make possible its de-

[13] San Diego *Union,* Aug. 25, 1870.

[14] Ibid., June 6, 1872.

velopment both as commercial entrepôt and sanitarium, as local figures so often pointed out. The railroad was long in coming; the health seekers were not. Campaigns to lure both to San Diego became one of the chief pastimes of the inhabitants. In 1870 Dr. Henry Gibbons remarked:

> The inhabitants have secured a large stock of thermometers and pluviometers, and have become zealous meteorologists, and determined to demonstrate the unparalleled sanitary values of their growing burgh. Thus far San Diego has led the race [against Los Angeles] and presents the strongest inducements to valetudinarians.[15]

News traveled even to New Orleans, where the Crescent City learned that "There is no doubt it [San Diego] will become a resort for valetudinarians as a Spa, and with outlay will be made famous beyond any on the Atlantic board."[16]

To fulfill such predictions, Gunn advocated the building of cottages and hotels for the incoming sick. In 1871 there were fourteen hotels, but only two were first-class establishments. Alonzo E. Horton built the city's first fine hotel and became the "Father of San Diego," the founder of a permanent New Town. After his arrival in 1867 Horton noticed that his cough had left him after six months. Foreseeing the region's possibilities for health seekers, he erected his hostelry, which for a time after its completion in October 1870 was the largest in southern California. Visiting the place in 1873, William I. Kip, Episcopal bishop of San Francisco, remarked, "With its broad airy halls and sunny rooms, it furnishes exactly what they [the invalids] need, and might adopt for its name one of the queer titles which the Chinese in their own country, bestow upon some of their hotels—the 'Hotel of Accomplished Wishes.' "[17] Throughout a generation, all of San Diego's hotel men catered to the sick in their advertisements, for the group was a large minority of the summer visitors and quite often constituted a majority of new arrivals during the winter months.

[15]Henry Gibbons, Sr., M.D., "Where Shall We Send Our Consumptive Patients?" *Pacific Medical Journal*, N.S., IV (1870), 312-313.

[16]San Diego *Union*, Mar. 28, 1873. John C. Daly called San Diego "the fabled Elixir of life if such does exist." *California Farmer* (San Francisco), Feb. 15, 1877.

[17]John Erastus Lester, *The Atlantic to the Pacific* (London, 1873), p. 211.

Winter housing was an acute problem, for private houses were not built fast enough during the seventies. Kip had noticed the San Diego-bound steamers filled with hopeful consumptives, while another visitor of the time described the hardships of his fellow invalids on shipboard. Thus San Diego experienced its first inflation. Mortality rates were inflated, too, since 40 per cent of all deaths in the community were due to tuberculosis. In August 1880 two medical investigators recorded, "It is doubtful whether there can be found elsewhere a population more extensively contaminated with the tendency to pulmonary disease."[18] With some exaggeration the resort booster Ben C. Truman wrote that the bulk of the population was composed of invalids. As the plight of health seekers was in most cases dramatic and thus noteworthy, everyone was conscious of their presence, and as a result their numbers may have been exaggerated to some extent by observers. In any case, after 1880 the proportion of sick to healthy people began to decline in San Diego. Better transportation and greater agricultural and commercial development appealed to a cross section of American migrants. The tourist trade had also begun by then. It was estimated that now only one newcomer in twenty was a genuine invalid. This may have been a minimization, as earlier figures were overestimates, for boosters of later days disliked to believe that only health seekers could discover the greatness of their San Diego.

One day in 1884 an extraordinary health seeker arrived in the celebrated city of "Bay 'n Climate." This was Elisha S. Babcock, who had recently heard that the sunny region was a sick man's paradise, and envisioning the possibilities of the Coronado peninsula commercially fed by rail connections with the bay, he planned the world's largest hotel and a new city. After Babcock had sold over a million dollars worth of Coronado lots and undertaken an extensive advertising campaign, the Hotel del Coronado materialized. When other land companies were depressed after the recent disastrous boom, Babcock's advertisement occupied prominent places in the newspapers of the San Diego region:

[18] Henry Gibbons, Sr., M.D., "The Inheritance of Pulmonary Disease: Its Possible Eradication, with Especial Reference to the Climate of San Diego," *Pacific Medical Journal*, XXIII (1881), 403-404.

> The Coronado Beach Comp'ny wants the Citizens of San Diego to live on Coronado Beach, believing It Is Much Healthier besides being more Pleasant. . . . Healthier—Because it is cooler in summer and warmer in winter. . . . Healthier—Because Coronado has the finest water in the world. . . . Healthier—Because of a better sewer system and the impossibility of the ocean breeze blowing the sewer gas from the sewages back to the homes of Coronado. Healthier—Because there is no dust in summer and no wind in winter.[19]

Local realty promoters and financiers connected health and community progress as they had elsewhere in southern California. In time, minor as well as upper-bracket employees of many banks and stores were mostly cured tuberculars. In fact, every occupation had a large representation of recently recovered men. The semi-invalid teacher was a familiar classroom character—and a health menace to the young. But consumption was not the only ailment; every disease associated with southern California health seeking—asthma, malarial fever, rheumatism, and various lung complaints—was found in San Diego. In this port town sea life helped many, or at least seemed to do so, and one might encounter on any street men who claimed to have recovered while working on the Pacific Mail Steamship Company's vessels and who now called San Diego their home.

The decade of the nineties was an era of considerable sanitarium building. With the pioneer Dr. Remondino, Dr. Robert J. Gregg conducted a sanitarium in downtown San Diego. Having come west for health in the eighties, William A. Edwards opened a sanitarium on the crest of Florence Heights and was a leading community figure by 1890. During the decade, the United States Army began sending tuberculous soldiers for treatment at the San Diego military establishment. Medical officers there continually reported favorable results, while navy men were suggesting that a marine hospital be built to take advantage of the curative effects of the climate. Despite the large tuberculosis mortality, San Diego's health officer reported a lower death rate than was found in

[19] San Diego *Union*, Oct. 13, 1889. Covering much of the front page of the San Diego *Sun* in repeated numbers late in 1889 was: "Coronado: The Ideal Climate. The Unrivalled Sanitarium of Europe and America!"

269 foreign and American cities.[20] By this time, San Diego had at last become a city. It had two keys to its destiny, a great bay and an extraordinary climate. Without these, the pioneer contingent of health seekers would never have come, to be followed by tourists and numerous industries peculiarly fitted to the area. The invalid did not create modern San Diego, but he certainly played a significant role in giving the city its initial boost. As in a chain reaction, the original momentum had not been wasted.

Detractors have traditionally claimed that San Diego has no back country. For health seekers, this statement was never true; even in the earliest period of their arrival they began pushing into the valleys north, south, and east of the urban center. About 1870 the ranchers of El Cajon Valley began to house several dozen boarders from town. As the air was drier and warmer inland than in San Diego, the number of sickly visitors increased. One settler, J. O. Minor, accommodated twenty-five persons. Becoming strange predecessors of the dude ranchers, other farmers of that region slowly increased their income with this type of guest.

During the boom of the eighties, Poway seemed about to become a significant interior resort, for homes were built rapidly and a two-story hotel rose there. Nevertheless, after more than seventy years the once promising village has barely attained a population of 300. Still, health seekers did found and permanently develop many surprisingly out-of-the-way communities in San Diego County.

Santa Barbara was the perfect example of successful advertising applied to a potential health resort. Until the late sixties the town had remained a pleasant Mexican village, a gathering of adobe dwellings near California's best-preserved mission. Yet, so picturesque a town inevitably attracted a few strangers, and as early as February 1867 Brigadier General James F. Rusling reported:

> The buildings were mostly adobe, of course, and all quite old; but the town had an appearance of comfort and respectability, if not of thrift,

[20] San Diego *Union*, Apr. 6, 1890.

> and the few Americans we met were sanguine of its future. . . . As a sanatarium, Santa Barbara was already being much resorted to by invalids, and doubtless will become more so when better known.[21]

As the seventies began, J. A. Johnson, editor of the Santa Barbara *Press*, rejoiced in the occasional accounts of recovery from tuberculosis of eastern invalids stopping there. Why not advertise the climatic blessings nationally, he thought. Johnson was an unusual man for his day. Both dauntless and untiring, he hoped to propagandize a million people. To him, southern California was not only a geographical unit but also an economic one, so he concluded that he would serve his community best by advertising the whole area from Point Conception to the Mexican border. Touring America to accomplish this purpose, Johnson emphasized the booster term "New Italy" and discovered that his region was still as little known to the average intelligent New Yorker as mid-Africa, then being publicized by Henry M. Stanley's adventures. Here then was an obvious challenge, and Johnson reveled in this crusade to inform a sickly world of hope from the Far West.

Results of Johnson's tour were far better than could reasonably have been expected. At one time crowds, gathered at the San Francisco wharf, had to be held back by the police when southbound ships sailed. For a time, a veritable human inundation had come upon Santa Barbara, and Johnson was its prime mover. Heralding the phenomenon and its creator, *La Cronica*'s publisher announced from Los Angeles:

> I will say here of Mr. J. A. Johnson that were it not for him and his indefatigable efforts, Santa Barbara would not to-day enjoy the prosperity she does. His untiring energy and perseverance in bringing before the world the advantages offered by this city and its climate to invalids especially entitle him to the gratitude of the entire community.[22]

[21] James F. Rusling, *The Great West and Pacific Coast* (New York, 1877), p. 324. Seeking the perfect resort, Dr. Thomas M. Logan declared it "the land of promise to the weary and desponding invalid . . . all prerequisites of health are to be found in measures so profuse that I would be accused of poetic extravagance were they duly portrayed." E. N. Wood, *Guide to Santa Barbara, Town and County* (Santa Barbara, 1872), pp. 36-37.

[22] Translated in Los Angeles *Herald*, Sept. 10, 1874.

Yet, the San Diego *World* editorially sneered, "A paper *bizarre* in the extreme is published in that home for invalids, whose ill-fortune has kept them ignorant of the superior climate of San Diego."[23]

At the time it was not uncommon for nearly a hundred health seekers to be landed by a steamer. Whenever this happened, men walked the streets all night unable to find lodgings. As a result, probably a majority of them soon left for less crowded resorts. In 1872 Santa Barbara, a town of 3,000, was first shaken by social changes the repercussions of which were comparable to one of the community's infrequent earthquakes. Joyously, Editor Johnson wrote:

> The town of Santa Barbara is full of strangers. About forty passengers came ashore on Tuesday last, and found it difficult to secure lodging anywhere. It will take a dozen hotels to meet the crowds of visitors when once the real character of Santa Barbara as a resort for health and pleasure becomes fully known. It has been a marvel to us for years that so few people could be convinced that the claims of Santa Barbara were absolutely genuine and well founded; but at last people are beginning to find its claims all good.[24]

For the newcomers a civic committee tried to secure all the spare rooms, while a company was hastily organized to build a temporary hotel—a large cookhouse in the center of a tract, surrounded by small cottages for families who reached the communal kitchen by plank walks. The town looked much like a community struggling to establish order after a natural catastrophe. As a result, residents forgot their accustomed suspicion of strangers and welcomed the ailing into their own homes.

During a period in which so many dramatic events were taking place this town-turned-sanitarium inevitably had its color and comedy. From the decks of incoming ships invalids saw as their first symbol of the promised land a large sign recommending "Metallic Coffins." Even a decade later the general confusion at the landing remained, its psychology hardly more comforting:

[23] Jan. 7, 1873.

[24] Santa Barbara *Press,* Mar. 23, 1872.

> As the steamer approaches town a bell on the wharf rings out its peals; then the denizens of the place rush to the wharf. A lusty fellow yells out, "Another carload of consumption." Doctors, with hawk like eyes, look anxiously for the poor victims that may fall into their hands. One of these gentry looked at me as if he would like to feel my pulse. I thrust it into the pocket of my ulster, and gave him a look which plainly said, "No you don't!"[25]

Critics attacked Santa Barbara's most obvious weak spot, housing. One disappointed paying guest said, "Santa Barbara is about a third-class hospital, with meagre accommodation at the highest rates."[26] At the time, he was right. Another complained about the poor quality of eggs and the lack of fresh vegetables, concluding, "the patient who tries to starve it out there will suffer more from hunger than most persons do from disease."[27]

Promoters realized that a really fine hotel would develop Santa Barbara's invalid trade to its full capacity. Consequently, a group of capitalists built the Arlington, famous throughout the West for a generation. Many of its patrons were well-to-do eastern invalids. The Arlington's grounds were beautiful, the furnishings expensive and tastefully chosen, and the service unusual for the region. Charles Nordhoff, virtual king of the boosters, enthused, "as there is but one Santa Barbara in the world, so there is but one Arlington."[28] Even the incomparable Arlington Hotel had its problems, for during the winter months guests might be turned away by the hundreds, while in summertime the hotel's patrons were pitifully few. Like the numerous stores and saloons which began to appear out of all proportion to the resident population, the hotel lost money during the warm months when tuberculars returned east or went to San Francisco, and Santa Barbara's seaside popularity failed to make up for the loss.

[25] T. H. Merry in *Pacific Rural Press* (San Francisco), Sept. 29, 1883.

[26] Quoted in San Diego *World*, May 4, 1873. According to one, the formula was that health seekers would live off the climate and Barbarenos would live off the health seekers.

[27] San Diego *Union*, Feb. 5, 1874.

[28] Santa Barbara *Independent*, Nov. 11, 1895.

Although in time there were rates and styles provided for everyone, Santa Barbara became primarily the refuge for the rich and retired. Its whole economy was molded by this factor and probably would have collapsed without health as a mainstay. By 1898 its role was summed up as follows, "A resort for the wealthy and unhealthy is all Santa Barbara aspires to now; and it is not disappointed; for hither these come from all parts; even from beyond the seas; a pleasant place wherein to live or die."[29] Real-estate values skyrocketed for the first time when the invalid invasion stimulated the sale of lots and farms. The health seekers aided business in other ways. Wealthy invalids built mansions, thus increasing the value of adjoining properties. In fact, from 1872 to 1875 property trebled in value. The nation was slowly recovering from the panic of 1873, and as it did so, more money flowed into Santa Barbara, a town which had never known the depression.[30]

Nevertheless, the health-resort bubble had grown too big. Seeing the invalid trade as a means toward a much more magnificent destiny, capitalists speculated. They hoped to develop their town as a great commercial port, and, as a contemporary reported:

> In 1874 the bubble of expectation was full-blown. Vandalism was rampant. Picturesque cottages were torn down and new buildings erected in their place. The town was looked upon as a future metropolis, and real estate commanded fabulous prices. Then the bubble burst, and hearts grew sad. . . . To-day Santa Barbara has accepted its alternative and has laid siege, together with Nice, Mentone, and Newport, to the distinction of being a sanitarium and popular resort.[31]

The "bust" was merely a recession, not a permanent setback, yet it was great enough to frighten the citizenry. A smallpox epidemic in 1878-79 in this "germ vacuum," as Dr. Brinkerhoff had called Santa

[29]Thaddeus S. Kenderdine, *California Revisited, 1858-1897* (Newtown, Pa., 1898), p. 216.

[30]Between 1868 and 1872 Santa Barbara County's indebtedness had been halved and its tax rate reduced. County population doubled to nearly 10,000. From 1870 to 1872 the town had increased about 600, due to the health rush. Wood, p. 20; San Diego *World*, Jan. 8, 1874.

[31]Edwards Roberts, *Santa Barbara and around There* (Boston, 1886), p. 17.

Barbara, was barely prevented. The outbreak certainly did nothing to restore wavering confidence. One critic in 1881 maintained, "I don't know a place in the United States where it is easier to do it [get accommodations] now. Houses that rented for $500 a year, now rent for $100, or more likely, stand empty."[32]

Sociologically, Santa Barbara benefited greatly from the coming of the invalids, many of them highly educated Yankees. These people were of "the better class," refined, traveled, and usually well off financially. A resident of years boasted:

> One thing is commonly remarked of them [the Barbarenos], their often high standard of culture. Graduates of our Eastern colleges, who in New England would crowd the ranks of lawyers and doctors . . . here give themselves to the quiet pursuit of agriculture. . . . In some valley, experimenting with the orange or walnut or olive, you shall find one who speaks of Yale or Amherst as his Alma Mater.[33]

Most of the southern California promotional tours began in Boston. Phillips' excursions left the Hub each week, as did Judson's and Carter's. All of them easily connected with Santa Barbara's transportation facilities. Explaining the phenomenon of the invalid elite, Josephine Sanford said that newcomers in Santa Barbara were usually of a superior character to those who came just to make money, and a good number remained as settlers. Thus, fitting easily into the homogeneous society already in existence, the health seekers built for the future and contributed to the "genial undercurrent of music, tea and card parties."[34] A more complete analysis would add that only the wealthy or those endowed with superiority in some field could survive economically as well as physically in this renowned health resort. Unlike larger Los Angeles, where more "plebian" invalids congregated, Santa Barbara provided almost no livelihood for such people. The town was small, jobs were

[32] Lewis, p. 290.

[33] Rev. Abraham W. Jackson, *Barbariana: or Scenery, Climate, Soils and Social Conditions of Santa Barbara City and County, California* (San Francisco, 1888), p. 17.

[34] *All about Santa Barbara*, p. 25.

few, real estate was high, and after the short-lived recession prices rose higher. As Los Angeles was supplied much earlier with cheap land transportation, that big city became an unwilling magnet for the East's poor sick. Santa Barbara was an outpost. Nothing but health was offered as its stock in trade.

Health remained the keystone of Santa Barbara's delicately balanced arch of hope. To disturb the residents one needed only to exclaim, "The reputation of Santa Barbara as a health resort is seriously threatened." Mayor George W. Coffin used this as his rallying cry in 1885 to get a better sewerage system.

Suicide, a notable aftermath of the quest for health, was notable in most southern California towns but became an especially lamentable problem in Santa Barbara. In 1893 the city's rate was 30 per 100,000, and except for San Francisco and Saxony this was probably higher than anywhere else in the world. At the same time, tuberculosis carried off one third of all who died in Santa Barbara, and it was claimed that the suicides were mostly tuberculars who did not wait to die a natural death. In the scenic surroundings of Santa Barbara, the invalid appeared to strike the only tragic note.

One can partly account for Santa Barbara's lasting popularity by referring to the splendid facilities for good hunting, fishing, hiking, climbing, swimming, and riding in the near-by countryside. The dangerously ill rode there, even if they had never done so before. There were "many very curious figures to be seen in the saddle, fit studies for Nast [the great American cartoonist of that day]."[35]

By 1900 Santa Barbara was settling into its destined role, which the health-resort boom had only fulfilled with unexpected swiftness. Critics insisted that the town was hedged in forever from greatness by mountains and an almost harborless sea; with more merit they asserted that the region was too damp for the cure of lung diseases. Yet, if faith can build, it had done so here.

Not far from Santa Barbara lies Ojai Valley, which Charles Nordhoff so effectively publicized. A resort town founded because of his nation-

[35] Santa Barbara *Independent*, May 12, 1885.

wide advertising bore his name. Laid out in 1874, Nordhoff consisted chiefly of two well-patronized invalids' hotels, the Ojai Valley House and Oak Glen Cottages. Although the continuous migration of health seekers to the area was modest in size, compared to the arrivals of the sick in other parts of southern California, a visitor of 1885 reported that not a house in the valley was without at least one person seeking relief from the ravages of tuberculosis or the discomforts of asthma. The valley had been opened up by health seekers who had faithfully followed the advice of a book, Nordhoff's celebrated guidebook. Since that day the region has continued to be a well-patronized resort for the well-to-do, if not necessarily the invalid.

Before the rise of the health resorts the hinterlands of southern California had barely been settled. To an important extent, the health rush changed this. An example of colonization by the sickly was Ontario. In the vicinity of that irrigated farming community one might find several levels of altitude which were sought by sufferers from lung ailments. Most popular of all was near-by San Antonio Canyon, which by the eighties had become a retreat for asthmatics, mostly easterners with a good accumulation of wealth. As early as November 1880 Dr. Joseph P. Widney had advocated the canyon as the ideal location for either a tuberculosis or an asthma hospital. Thus the scheme was older than Ontario itself. In 1891 prosperous Ontarians and neighboring citizens of Pomona met to organize a sanitarium company. "It is the only scheme ever proposed whereby a man can lay up treasures in heaven and in his own pocket at the same time," philosophized the local press.[36] Eventually, sanitariums were built, but none ever achieved either the pretentious dimensions or the financial rewards envisioned by early promoters. Undaunted, health seekers economically stimulated the region even without such ideal provisions.

Farther east, San Bernardino had a good head start as a sanitarium site. In 1872 Nordhoff had mentioned it in his popular book, and like other places his boosting had blessed, the community soon was welcoming a daily arrival of letters asking about the valley in which it was lo-

[36] Ontario *Observer*, July 18, 1891.

cated. Here, again, ordinary amenities found in the East were poor or unknown in a frontier area unused to any but a few hurried visitors. Those whom publicity had brought, the dry climate would retain. In this way inland southern California gained a large share of its early population.

The year before Nordhoff's book appeared, a Baltimore invalid came to board with Myron H. Crafts, a pioneer in San Bernardino Valley. Other health seekers followed, and Crafts had to enlarge his ranch house, adding a second story and turning his place into a "health hotel." Crafton Retreat was well known for over a generation, and in time it became the nucleus of a new town, Crafton. Feeling a founder's responsibility to the community he had established, Crafts built an Indian trading post, a school, and a church; in 1882 the Crafton School District was organized.

San Bernardino and the outlying countryside soon were dotted with crude sanitarium-hotels. A few of these followed the course that Crafton Retreat had so successfully pioneered and helped the region progress. Unfortunately, a larger number preserved less happy memories for their patrons. An alert newspaperman of the day gives us perhaps the best description ever written of a jerry-built sanitarium in early southern California. Having arrived at San Bernardino in 1876, George F. Weeks put up at a crude resort in the foothills:

> The Sanitarium was a typical California farmhouse, two stories in height, with porch extending entirely across the front. A long extension had been constructed in the rear, the upper portion of which had a hallway leading down the center. Cubicles opened at regular intervals on either side, each roomlet being large enough to accommodate one half-size bed, one chair, one washstand, one small bureau, one set of hooks for hanging clothes, one tallow candle, one emaciated invalid of the scarecrow style, of which I was said to be a type.[37]

About 90 per cent of the patients had infected lungs; everyone inhaled the breath of others, for since drafts were considered fatal, venti-

[37] George F. Weeks, *California Copy* (Washington, D.C., 1928), p. 48.

lation was extremely poor. Most unforgettable of all the torments of the incarcerated was the constant chorus of coughing from 9 p.m. until morning. First one occupant would have a paroxysm, then another. When the sun rose, all gathered around the sitting room fire, each telling in detail of his bad night. Routine, the curse avoided by later sanitariums, was endless, and the isolated spot received little news of a healthier outside. If the day turned bad, the patients remained at their morning stoveside. When it was sunny, they gratefully crept onto the porches or out onto the adjacent plain, an uninspiring mass of rocks and scattered brush, far in time and planning from the later studied elegance of well-landscaped sanitariums. Consequently, most patients simply awaited the end, and in that at least they were seldom disappointed.

Within a decade, however, San Bernardino County had a modern health resort, the Rabel Sanitarium. It specialized in mineral baths, as did most hospitals in that region. The dry air helped tuberculars while hot waters were said to improve the condition of asthmatics.

During the eighties, permanent invalid pilgrimages brought population to the area. The San Bernardino *Daily Index* noted in 1888:

> San Bernardino County like the rest of South California has been practically settled and made what it is by invalids. Today it is hard to find any family that has come into the county within the last fifteen years, but what had one or more members who were in poor health.[38]

Due to its protected valley, San Bernardino seemed to offer optimum conditions for tuberculars. So many had arrived by the eighties that it was observed that "the cough of the consumptive is heard" everywhere.

Redlands was a key community in this fertile inland region which came to be called the "Great Orange Belt and Sanitarium." When the settlement was less than a decade old, observers described it as "essentially a Connecticut colony," because the mayor was a health seeker from New Haven and other men of prominence had come from the same state, many of them for similar reasons.[39] Built by the sick, Redlands was logically

[38] Nov. 14, 1888.

[39] *Citrograph*, Apr. 26, 1890.

headquarters of the short-lived *Southern California Sanitarian and Climatologist,* a learned journal which attempted to advertise the area for health seekers. After 1900 "The Settlement," an excellent tent hospital, was widely hailed and supported by the private citizens of Redlands. It was administered by local charitable associations. Near-by Mentone, named for the famed health resort on the French Riviera, had a boarding house for tuberculars, but the town's trifling size discouraged as much settlement as Redlands was enjoying. The time when health seekers pioneered in building colonies and founding towns was relatively brief. In spite of obstacles, Redlands remained for several years their special haven in the inland region. In 1904, for instance, 90 of the locality's 169 deaths, or 53 per cent, were due to pulmonary tuberculosis.[40] Even in 1910 one could describe the tent communities typical of Redlands and San Bernardino thus:

> Then go where one will, he will come upon isolated tents, some inhabited by lonely sufferers, and some by whole families. There are tents in front yards and back yards, in vacant lots, by country roadsides, on farms and ranches, in secluded canyons, in deep recesses of the forests, far out upon the deserts, and away up in the mountains. They are the camps of the Arabs of the Southwest—a forlorn, homeless and almost hopeless multitude of wanderers, chasing the phantom, Health.[41]

Poised between desert and mountain heights, the Banning-Beaumont region enjoyed the benefits of both climatic zones. Coming during the eighties, its invalid founders turned to a variety of agricultural pursuits, with fruitgrowing predominant. From the beginning Banning was slightly ahead of Beaumont in population and development, and it has maintained that lead. Editorials and pamphlets of the Banning *Herald* helped to raise the proportion of sick residents until some observers mistakenly believed that invalidism was actually universal there.

Banning's importance was made possible by its good lodgings. Nev-

[40] C. H. Alden, "Some Southern California Health Resorts, 1904-5," *Transactions* of the American Climatological Association, XXI (1905), 48-49.

[41] John L. Cowan, "Climate and Consumption," *Overland Monthly,* 2nd Ser., LV (1910), 250.

ertheless, furnished places were scarce, since at an early date each dwelling was by city law thoroughly renovated after a tubercular vacated it. As the poor health seekers did not often frequent this upper-middle-class refuge, crowding and public charity never became great problems. Furthermore, there was little floating population in Banning, for most of the invalids had come to stay. A good drugstore and other mundane but indispensable facilities made their residence comfortable.

The area's leading physician, John C. King, was one of these who came because of poor health. A lover of statistics, he compiled many of them during his useful career. Of the first 200 tuberculars he treated, 78 were cured, 49 improved, and only 59 died; the remainder showed no change—this despite the fact that nearly all of them had come in the last stages of phthisis. This victory about halved the current death rate of tuberculosis. King had thoroughly investigated the region, his own particular Eden, and felt sure that the 2,000-foot elevation, its "aseptic atmosphere," good natural drainage, and wholesome mountain water were responsible.[42]

Still collecting data in 1906, King published for academic circles his most exact statistics, and from them we get a rare thing, definite figures on the health-seeking migration, as precise for one area at least as any could ever be. By now a resident of twenty-two years, he had canvassed every house and family in Banning, a town of 1,000 population. He found that actual tuberculars constituted but 12 per cent of the citizenry, although 53 per cent of the resident families had come to the area suffering from tuberculosis or because at least one member of the family was a consumptive. Consequently, 31 per cent of the dwellings were, or had been, hotels or boarding houses where tuberculars lived, while 72 per cent of all homes had housed such patients. Due to zealous local physicians, the populace had been carefully instructed in the problems of tuberculosis. As a result, in 1906 there was not a single case which had developed locally. The Tudor House, where King was medical supervisor, was devoted exclusively to tuberculars. Once a room there was vacated, sealing and fumigation followed, and formal-

[42] Banning *Herald*, July 18, 1891.

dehyde was used for disinfection over a period of at least twenty-four hours. Any patient who expectorated anywhere but in his own receptacle was discharged regardless of his wealth or social position.[43]

With such a small population, Banning could remain in nearly test-tube perfection, a showcase health resort. Evidently, the number of actual tuberculars was smaller than the public thought. The influence of these unfortunate people throughout southern California had always been stressed and usually exaggerated. Of course, Banning's statistics refer only to actual uncured cases. Then, too, the community had a large group of asthmatics, rheumatics, and sufferers from bronchitis, not incapacitated but certainly health seekers. They were seldom played up in the statistics or mortality rates of other areas, for death was less frequent and contagion impossible in their cases. Beaumont played a secondary role in this story, although it was long advocated for asthmatics.

Man has always sought spiritual salvation in the desert. For physical well-being, too, he has looked to Nature for abundance in climes where it seemed nothing could be given. In desert California, far from the ocean and its atmospheric effects, tuberculars found the dry air that so often proved beneficial to them. At Oro Grande on the Mojave River a little colony of Angelenos settled in the early eighties. Several southern California doctors pioneered in that arid wilderness, although poor housing and faulty transportation, the ever-present check on wide-scale health seeking, prevented larger numbers from following their examples. By 1900, however, such places as Big Rock Villa on the edge of the Mojave Desert and near-by Antelope Valley offered modest facilities for asthmatic and pulmonary sufferers who valiantly tried to farm.

In 1873 Dr. J. P. Widney had first advocated that the Colorado Desert be exploited as a site for a great sanitarium. Unfortunately, the popular conception of deserts defeated him. Then, in 1886 Dr. Wellwood Murray and his wife established a small resort at Palm Springs where they regained their health and soon began to welcome dozens of

[43]John C. King, M.D., "A Review of What Has Been Done for the Prevention of the Spread of Tuberculosis in the State of California," *California and Western Medicine,* IV (1906), 166.

boarders, who were mostly tuberculars. A millionaire from San Francisco was among these paying guests. With recuperation, he unwisely assumed that he had regained permanent health, and left. A relapse brought him back; recuperation and relapse followed, with death the finale. Yet, his frequent sojourns at Palm Springs spread the good news about this oasis, conveniently near the populated areas of southern California. By 1899 the Murrays' resort had become the nucleus of a town. Their hotel, run according to their dietary beliefs, housed thirty invalids at that time. The founders built a church and schoolhouse for neighboring Indians. Less substantial structures rose near-by, more typical of that day's desert colonies; these were the little groups of tents set up by tuberculars. Inevitably, some of the hopeful settlers had their trusty cows staked outside.

In the Salton area the fortunate combination of low altitude and aridity was believed to facilitate breathing. Because of this, several dozen rheumatics and tuberculars went there in the nineties, while at Needles, a key railroad point on the Colorado River, a colony of tuberculars put up with bad conditions in housing and poor medical facilities.

In a sense, the appeal to the sick to populate the desert was a futile crusade. Yet, desert settlements were made. Summarizing the reasons, W. B. Ewar said:

> A race is coming on—a race has already arrived—who are so far gone and effete that they require not so much exercise or recreation as absolute rest—rest, not only from work and the worry of business, but rest from the excitements and the excesses of a too active or a too luxurious life. If we could confine them in a vacuum, or temporarily suspend animation, it would be the proper treatment for them. It is because there is nothing or so little on these deserts that they would constitute the most suitable asylum for a class of patients whose vitality is so nearly extinct.[44]

Ewar had struck the truth. It was not the climate that cured, but the new way of life—rest, relaxation, a psychology of serenity—which saved

[44] *Mining and Scientific Press*, LVI (1888), 119.

many a health seeker. Others might spend all their money reaching and living in a pleasant southern California resort town and die because they refused to change their lifelong regimen.

During the next decades, Arizona and New Mexico had better success in making popular their arid lands. Thus the health quest in southern California had served only as a minor factor in America's desert development. For that matter, nonirrigated regions beyond the citrus belt were hardly developed by anyone until the twentieth century. Until recent and rather unexpected developments at mid-century, southeastern California remained, with a few outstanding exceptions, a wilderness, in striking contrast to the coveted lands to the west.

Although its achievement in the peopling of southern California was uniform neither in geographical distribution nor in chronological development, the health rush had crowded the chief resort towns and, to some extent, lesser communities. New villages were founded on the basis of a promise of health. Many aspects of urban existence were modified. Transportation to outlying regions thus became a necessity. All in all, the health seekers well deserved the title of town builders, if not empire builders.

CHAPTER V

Medicine and Pseudo-Medicine

THE INVALID's best friend, his doctor, influenced the search for health in California in more than a mere medical sense. The medical man's very numbers made obvious and systematic southern California's expanding role as a giant sanitarium. At the end of the generation-long health quest the situation was objectively summed up by statistics. In 1900 the United States census showed that California had more doctors in proportion to population than did any other state. At that time the national average was one to 655 persons, while in California the ratio had reached one to 416. Even Colorado, the favorite retreat for invalids preferring the Rockies, had a less imposing record, one physician for every 462 residents, but the Centennial State had not as yet become so densely settled a health-resort area as had southern California. Commenting facetiously on local conditions, a Los Angeles *Times* columnist said, "Whether the large number of doctors in California is due to the great number of invalids, or whether the large number of invalids is to some extent due to the great number of doctors, is a conundrum which is not so easy to solve."[1]

It was appropriate that the press of Los Angeles should make such observations, for the city's medical fraternity was primarily responsible for the astounding figures. In 1901, according to the *Los Angeles City Directory*, there were 375 physicians and surgeons, or one for every

[1]Los Angeles *Times*, Illustrated Magazine Section, July 21, 1901, p. 25.

273 Angelenos.[2] This included only those who stayed long enough to be counted. Traveling practitioners and the endless crowds of quacks were never enumerated.

Even in the pioneering days of the health migration Angelenos became conscious of their community's surplus of doctors. Dr. G. W. Linton declared in 1875, "We have a larger number of doctors in this city than any city I know of in proportion to the population."[3] As that population grew and life became more sedentary, this trend was even more pronounced. By 1895 J. A. Munk, M.D., was not surprised that the census of 1890 gave Los Angeles a larger percentage of physicians to total population than any other city on earth, a high of less than 200 inhabitants per doctor. With good reason, he predicted that "others are coming and more are bound to come—there are a dozen eclectic physicians and all are seemingly doing a satisfactory business."[4]

When in July of the same year the *Los Angeles Polyclinic,* a medical journal, published its first issue, its chief reason for being was formally announced as follows:

> The City of Los Angeles with its rapidly growing population of nearly one hundred thousand, and its reputation as a sanitarium, affords peculiar facilities and advantages for the publication of a high class medical journal. It is a fact beyond dispute that owing to the great influx of invalids "The City of Los Angeles" has a greater number of medical men than any place of its size in the world.[5]

Writing in 1897, two Angelenos described the now alarming conditions: "The Medical Fraternity of Los Angeles, including all who practice medicine here—of all schools or of no schools—will reach in number beyond four hundred," all of which they credited to the influx of

[2] *Los Angeles City Directory, 1901* (Los Angeles, 1901), pp. 1320-1322. Yet with a population of 102,489 in 1900, Los Angeles had only 12 hospitals. Ibid., p. 1286.

[3] Los Angeles *Express,* May 7, 1875.

[4] J. A. Munk, "Impressions of Southern California," *California Medical Journal,* XVI (1895), 175-176.

[5] *Los Angeles Polyclinic,* I (1895), vi.

physicians as health seekers.[6] So bad was the situation by 1908 that Dr. H. S. Delamere warned his colleagues that Los Angeles County had three to six physicians where only one was needed, and yet more came. The year before, the county medical register had shown there were 741 doctors.[7] Equally a result of the migration of invalids was Pasadena's list of a dozen physicians in 1893, increased to 95 by 1900.[8] With less than 6,000 people in 1891, Riverside had 13 medical men advertising in its press.[9]

As was mentioned earlier, numerous doctors had urged health seekers to go to California. In addition, many made the trip themselves for their health. A good number of southern California's most famous health seekers were members of Galen's profession. Speaking before a medical gathering in 1886, Dr. Walter Lindley observed:

> It would take hours to report all the cases of physicians who have themselves come to Los Angeles and other points in Southern California suffering from phthisis of various types and stages, and who are now following their chosen profession in apparent good health.[10]

Such doctors were most fervent, and usually highly successful, in persuading others to come west. While living in his native Ohio, Joseph P. Widney had determined to become a doctor, but poor health occasioned his move to California, where he finally achieved his desire by earning his doctorate. Widney opened an office in Los Angeles, and in his long lifetime he witnessed the most important decades of the health quest. Newspapers, magazines, and several volumes of books preserve his name and thoughts on the healthful climate. His achievements for southern California and the local fame he won were better evidence than

[6] Atchison and Eshelman, *Los Angeles Then and Now* (Los Angeles, 1897), p. 68.

[7] H. S. Delamere, M.D., "The Over-Production of Doctors," *California and Western Medicine*, VI (1908), 270.

[8] *Pasadena City Directory, 1893-4* (Los Angeles, 1893), p. 87; *The Moore Pasadena City Directory, 1900* (Pasadena, 1900), pp. 292-293.

[9] Riverside *Press*, July 11, 1891. Chino, needing a doctor, complained that a near-by town with a population of 4,000 (Ontario?) had 22, while another boasted six physicians for only 1,200 potential patients. Chino *Valley Champion*, Dec. 25, 1891.

[10] Los Angeles *Tribune*, Nov. 8, 1886.

even his great mass of written words that the climate should never be underestimated. As a founder and trustee of the University of Southern California, Widney helped to expand the institution and served as dean of its medical school, which he had established. It was natural that he should also lead in the organization of the Los Angeles Medical Association. He boosted the area for over sixty years. Blind, deaf, yet constantly active, even as a nonagenarian Widney persisted in his unquenchable optimism.

Chicago's loss and one of southern California's greatest medical gains was Norman Bridge, prominent physician and instructor at Rush Medical School. Bridge had visited Los Angeles with sick friends in the eighties and was so impressed by the climate that upon his return home he unreservedly recommended the region to his tubercular patients. In 1891 he was forced to take his own advice. There is a story, probably true, that Bridge used his own sputum in a tubercle slide test alongside another acknowledged to be an infected specimen and thus discovered that he had the disease. Bridge became an authority on tuberculosis and a prolific writer on medical and philosophical subjects. Through his connection with Edward L. Doheny and the oil industry, he made several million dollars. At his death in 1925 he left a considerable bequest to the Southwest Museum, of which he had been president. Bridge's medical activities connected him closely with the University of Southern California, although he also gave liberally to other educational institutions in the area, particularly the California Institute of Technology.

As a rather gregarious man, Dr. Bridge found a large number of his kind in Los Angeles, doctors who had come to the Pacific Coast because of tuberculosis. Among them were Titian J. Coffey, an obstetrician, Albert Soiland, the Swedish cancer specialist, George L. Cole, C. Corey, and Sumner J. Quint. The last-mentioned thought it a good idea to organize the group into a unique club, and so at the century's end the "T. B. Club" was founded. For more than twenty years it continued for social purposes, holding occasional meetings and at last gradually dying away. This odd fraternity, however, did not cease because its members died of tuberculosis. On the contrary, all lived long and useful lives except Dr. Corey, who succumbed to the disease shortly after the group had been organized.[11]

Sumner J. Quint had come from Massachusetts as a youthful novice in western ways, but he learned about the new land literally from the grass roots by tramping through Ojai Valley in search of a cure for tuberculosis. Having achieved his ambition through outdoor living, he satisfied another desire by becoming a doctor.

Other health-seeking physicians helped raise the caliber of local medicine above what otherwise could have been expected of the Far West in the eighties and nineties. Among these was Dr. Stanley Black, a well-known pathologist from Chicago, whom Dr. George L. Cole had persuaded to "try Los Angeles." In 1905 Black was president of the local medical association. Several members of the University of Southern California medical faculty had been tuberculars; indeed, they were in effect another "T. B. Club," though never so organized.[12] One of their students was Dr. Joseph Knopf. Influenced by tuberculosis and its problems, which he could hardly have ignored in the Los Angeles of that day, Knopf later studied in Europe, and his thesis at the University of Paris, "Sanatoria and Tuberculosis," won the international prize of the Tuberculosis Congress at Berlin in 1900.

San Diego had its share of health-seeking doctors, the most famous of whom was Peter C. Remondino, who had come from another health resort, Minnesota. By 1875 he was city physician and in time first president of the city board of health, a position Remondino held until 1921. Medical climatology, particularly localized in southern California, was his lifelong interest.

An uncounted number of mediocre doctors came west for various reasons of health and sought a precarious livelihood by conducting minor mineral spring resorts or primitive farmhouse sanitariums. Others followed the admonition of George Wharton James, who told the Southern California Medical Association to write articles for eastern papers boosting the local health resorts, and:

[11] Much of this information from interview with Dr. Sumner J. "Pat" Quint at Los Angeles, June 21, 1951.

[12] Interview with Dr. Raymond G. Taylor, Los Angeles, June 19, 1951.

> The result will be the placing before hundreds of thousands of people, in the course of a few months, of an array of facts in regard to our climatic gifts that the weak and infirm, the aged and the invalid . . . will be induced to further investigate . . . be restored to health, and thus give to us new settlers. . . .[13]

As we have seen, doctors had been doing this for a generation. T. C. Stockton of San Diego published tuberculosis statistics for out-of-state physicians to study, while J. F. T. Jenkins of Riverside wrote papers upon his region's advantages for asthmatics; these articles by Jenkins were published in the Canadian medical press. Meanwhile, numerous doctors lectured their local medical societies on maladies particularly aided by the western climate. So enthusiastic had some become by the nineties that southern California's booster-born epithet, the "Italy of America," was renounced; Italy ought to be called the "Southern California of Europe."[14]

In the relatively new field of medical philanthropy many doctors were to be found actively helping the sick. One summer in the late nineties Walter Lindley and his colleague Dr. F. T. Bicknell went high into the San Jacinto Mountains for a long-deserved vacation. There, Lindley got an idea, a plan to establish in this beautiful highland a tuberculosis sanitarium. Giving up their vacation plans, the two men hurriedly took options on several dozen acres in the area they had visited. Almost universally the physicians of Los Angeles approved the scheme and joined to establish the hospital.

With this beginning, in Strawberry Valley, southeast of Banning, the Idyllwild Sanatorium became a fact; it was opened on April 10, 1901. Behind that inauguration were the investments of forty prominent local physicians and twenty other important citizens organized as the California Health Resort Company with a capital stock of $250,000.[15] Spurning profits, the doctors aimed to get invalids out of the hotels and board-

[13] Pasadena *Daily Evening Star*, Dec. 5, 1894.

[14] K. D. Shugart, M.D., "Climatology and Hygiene," *Southern California Practitioner*, VI (1891), 287-296. Some local genius, paraphrasing the famous Italian saying, "See Naples and die," proclaimed, "See California and live."

[15] Interview with Dr. Quint.

ing houses of large towns and into the sunshine and mountain air. As such an institution as they had conceived of would be ineffective if dedicated to those far-gone in tuberculosis, only persons in the early stages of the disease were admitted.

The company adopted the cottage plan of layout, with a large central building containing dining room, solarium, parlor, offices, kitchen, forty bedrooms, and several private baths. No monotony plagued this model resort, for a gymnasium, golf links, and riding stables were constructed, and mountain guides were furnished the fortunate patients.

Yet, Idyllwild failed. The altitude, 5,250 feet, was too high for many invalids. More important, its inaccessibility kept the region from being genuinely popular. About 1907 a fire wiped out the main structure, and the company's deficit mounted to $125,000. The tract was finally exchanged for real estate elsewhere, and the idealistic investors managed to "break even."

Idyllwild was not unique; its counterpart, Loma Linda, was advanced by the foremost doctors of the Pacific Hospital. Interestingly enough, this resort, four miles west of Redlands, was conceived in much the same manner as Idyllwild. The investors, a hundred leading citizens of southern California, included eighty physicians. Unlike Idyllwild, Loma Linda took over a hotel already built. The structure had been put up in the midst of the great boom in 1886. Unfortunately, Loma Linda shared the fate of Idyllwild, and its quarters once more reverted to hotel status.

By the nineties southern California's medical men were at least as far advanced in the study of lung diseases as their eastern colleagues. Few individuals have written more words on tuberculosis than Francis M. Pottenger; not many have shown more interest in the disease than Norman Bridge; and probably none matched the philanthropy of Walter Jarvis Barlow.

As these men worked against disease, they had in their wake an army of quacks whose hindrances to medical progress were as numerous as the variety of microbes. As it had elsewhere, the caduceus, ancient symbol of the medical profession, had its counterfeit on the Pacific Coast. Unfortunately, in the case of southern California, the evil was multiplied. Historians Atchison and Eshelman wrote in the nineties that there was in Los Angeles

> A medical fraternity replete with men whose ability and standing range from that designated "best in the United States," to the very lowest of the worst quack-ridden city . . . and as the city is peopled by many health seekers, the quacks, irregulars and incapables fatten on the unfortunate.[16]

Paradoxically, the search for a restoration of health had caused this coming of best and worst. A few years later the editor of the *Christian Advocate* told a Methodist conference in Los Angeles, "Boston has the reputation of possessing more of the ilk [quacks] than any other city on earth, and I felt sure the 'Hub' was really in the lead as to numbers and variety—and she is, barring Los Angeles."[17] In 1905 a member of the local board of medical examiners called Los Angeles County the "Mecca of the quack," and few doctors would dispute the unpleasant nickname.[18]

Physicians did fight their perverted counterparts, even though the average citizen was not fully awake to the dangers of quackery. The Medical Practice Act, passed by the California legislature in 1876, was insufficient to stop malpractices. Unfortunately, although several times modified, it remained until 1901 an often ineffective statute. The president of the Los Angeles Medical Society blamed the people themselves for the successes of quackery. Yet, this was the golden age of the medicine show and an era without pure food laws, which were not finally passed by Congress until 1906. During that time the public was only partly educated in any sense and gullible in almost every way. Moreover, southern California held more desperate sufferers than any other American area of its size and population, and even in less promising lands hope has created more mirages than a world of deserts.

The public press was usually disgracefully willing to spread deceptive advertisements from which it profited. Journalists often published quack

[16] Atchison and Eshelman, pp. 68-70.

[17] Los Angeles *Times*, Illustrated Magazine Section, May 22, 1904, p. 27.

[18] "The Mecca of the Quack," *Los Angeles Medical Journal*, II (1905), 19-20.

testimonials, and some even gave their approval to supposedly enthusiastic descriptions by the paper's reporters. Fortunately, there were always a few exceptions. True to his character and honor, Scipio Craig, the man who in his journal crusaded for a sovereign state of South California, spurned the quacks and refused their copy. Not for the first time, he wrapped himself in indignation and editorialized:

> "An old physician, retired from practice," is hereby informed in most emphatic terms, that his advertisement "a receipt sent free for the cure of consumption" cannot be inserted in *The Citrograph* for any amount of money he may happen to have handy. While in the practice of pharmacy we have had more than one of these "free" prescriptions brought in to be filled. They were all bare-faced frauds. When we have to eke out a living by admitting such fake advertisements to our columns, this journal will be promptly offered for sale—50% off for cash, in order to quit business.[19]

Yet this was an all-too-rare reaction. Most newspapers were safe in accepting the quacks' literature, for these creatures were literally "fly-by-nights," usually remaining but a few days in any given town. Advance notices appeared in the press, generally accompanied by lengthy accounts of great victories over pain in the East or in other southern California communities far enough away not to be traced immediately.

In September 1875 Los Angeles welcomed one of these tricksters, who claimed that he had successfully performed at Anaheim, Santa Barbara, and other points in the "Sanitarium Belt." He offered to stay only two weeks but remained longer, always emphasizing his faultless services to the health seeker and warning him that it was "a mistaken idea for the invalid to rely entirely upon a change of climate to effect a cure."[20] If he had said no more than this truth, he might have been well remembered by posterity. Yet, this statement was only the prologue to his conclusion that it was not southern California but the practitioner whom the invalid needed.

[19] *Citrograph,* Feb. 8, 1890.

[20] Los Angeles *Express,* Nov. 18, 1875.

Even Santa Barbara attracted quacks. In 1883 a typical deceiver appeared in town. It was the practice of such people to use an alluring title, smacking more of the sideshow than the medical profession. This particular lady proclaimed herself "Queen of Magnetism" and offered to her listeners both electropathic and allopathic treatments for "all diseases no matter what name or nature, with never failing success."[21] Her specialties were an odd combination indeed—tuberculosis, venereal diseases, children's ills, St. Vitus dance, scrofula, and ulcers. To get everybody's business, she would also perform as a clairvoyant and spiritualist.

By the nineties medical vagabonds had become more colorful. Probably most dramatic of all was the so-called "Boy Phenomenon." Reputedly invited to Pasadena by some of its most prominent citizens, he arrived there "after many triumphs" in November 1894 and claimed to be able to cure more than a dozen maladies yielding to animal magnetism. The night of his first demonstration a reporter of the Pasadena *Star* went to the opera house to chronicle the moving event scheduled to occur there. On the scene were large numbers of health seekers at their most pitiful and worst, psychologically as well as physically:

> Indeed, so great was the rush that twice as many went as could by any possibility get into the house, and the down-pouring stream of humanity at half-past seven met an overflow equally as great returning, disappointed but good natured . . . never but once has such a sea of humanity greeted the gaze of him who looked back into the auditorium from the front boxes. Every nook and corner, gallery and all, was crowded, even to the aisles, and the boxes themselves were overrun by the crowd. . . . A corner at the right of the parquet had been reserved for such as desired treatment, and here were gathered such an array of cripples as Pasadena has never before seen.[22]

Rheumatics, considered ideal subjects for the treatment of animal magnetism, were most conspicuous. The poor were to be doctored free, the rest to pay "a small fee." During the next few days some cures were re-

[21] Santa Barbara *Independent,* Oct. 31, 1883.

[22] Pasadena *Daily Evening Star,* Nov. 14, 1894.

ported, especially among these rheumatics. Later, and characteristically too late, the healer was "exposed" by a local resident, A. J. McClatchie, who said that the only thing phenomenal about the whole incident was the youth's cheek. McClatchie had questioned all those treated and failed to find a single person who had been genuinely benefited. With common sense uncovering one maneuver of quackery, he reasoned, "if these men were really able to make the cures they pretend to, it would be unnecessary for them to move. They would have enough practice in such a town as Pasadena or Los Angeles to keep them there indefinitely."[23]

An integral part of quackery was the nostrum. Patent medicines, like a good many health foods made locally, disappeared as fast as they came on the market, and their literature was an experience to read. In 1883 the "Medical Triumph" came forth as a cure for asthma and other lung troubles. At $2.50 a bottle one could try a concoction called the "Temple of Health," a remedy—it was claimed—for epilepsy, asthma, Bright's disease, dropsy, rheumatism, and that vague Victorian complaint, dyspepsia. The Abbo Medical Institute "guaranteed" quick and permanent cure for most of the ills publicized by its competitors, with liver, kidney, stomach, and urinary troubles, rheumatism, neuralgia, and sciatica added for good measure.[24]

Sincere, but little more successful than the suspect concerns, was the Nixon Depurator Company's inhaling cabinet for the tuberculous. In 1895 it offered to cure consumption, catarrh, bronchitis, colds, and asthma. The patient would spend but ten minutes breathing the treated air, and then he might expect definite relief. Basing their device on the same principle as did the climate seekers and their doctors, the sponsors of this cabinet reasoned, "If there is any merit in inhaling the atmosphere in the glorious climate of Southern California, there must certainly be

[23] Ibid., Nov. 22, 1894.

[24] Medical terms were used more vaguely at this time than they are today, because of a lack of knowledge. "Rheumatism" was often synonymous with arthritis, although it is now known that there are about a hundred types each of both diseases. "Dyspepsia" is now called nervous indigestion. The term probably encompassed many poorly diagnosed gastro-intestinal complaints.

merit in inhaling atmosphere scientifically."[25] For those not helped by the western sunshine and air, a San Francisco gypsy offered to cure tuberculosis with a diet of dog meat.

With the enactment of vigorous federal laws as well as local health legislation in California, quackery began to decline. By the time the victory was certain, the health rush had diminished, too. All in all, the medical profession, with its great achievements against diseases most prevalent among health seekers, outbalanced the activities of quacks. Today, the region is famous throughout the world for its experts in pulmonary tuberculosis, a direct result of the early influx of invalids. Although quacks and doubtful healers of all kinds still congregate in southern California, they lack the perfect market of an earlier era, and their propaganda is hedged in by good government, an honorable press, and a mature and vigilant public opinion. The caduceus has become exalted, its counterfeit exposed.

[25] Los Angeles *Express*, Feb. 2, 1895.

CHAPTER VI

Faith and Charity

As THE nineteenth century neared its close and the health rush was reaching its peak, the Reverend Dwight L. Moody came to southern California on one of his last missionary peregrinations. One evening while he was preaching in Pasadena's Tabernacle, the famous evangelist looked about him at his hearers and said:

> How we cling to life. Thousands [are] here in California seeking health, bearing loneliness, homesickness, heartache and separation from dear ones, all that they may add a few years to life's span. Soon this audience will have passed from earth.[1]

Moody, himself soon to die, thought that the health seekers should be preparing themselves for the coming life rather than spending precious earthly time on their physical problems. Other ministers held various opinions on the issue. The Episcopal bishop and pioneer California preacher William I. Kip had noted in 1873:

> They come from every State on the Atlantic coast and in the West. Many of them are churchmen, and it is of great importance that we should have the right kind of clergyman to sympathize with them, and administer spiritual consolation in the hour of sickness and death.[2]

[1]Pasadena *Daily News*, Feb. 28, 1899.

[2]San Diego *Union*, June 1, 1873.

Many men of God had come west lured by the hope of life, and as semi-invalids or settlers who had been cured they understood well these human puzzles ever multiplying as the influx of invalids continued.

Faith offered consolation for the next world, but it might also give hope for relief in this life. Because of the latter factor, numerous faith cures developed in southern California. Most important of all religions influencing and being influenced by the health seekers was Christian Science. Offering hope where climate and medicine had failed, Christian Science was a natural conclusion for numerous sick persons. The result was an unusual growth of the denomination in southern California. Time, achievements, and the admissions of so persistent a critic as Mark Twain have shown that the movement has much more than a kernel of truth in its dogma. The connection of Christian Science with the California health quest seems inevitable from several passages in Mary Baker Eddy's famous work, *Science and Health, with Key to the Scriptures.* In reference to tuberculosis she wrote, "What if the belief is consumption? God is more to a man than his belief, and the less we acknowledge matter or its laws, the more immortality we possess."[3]

By the eighties this new faith had entered its pioneering missionary activity with energy, and this was the same moment that southern California had achieved national fame as a health resort. In 1887 Emma S. Davis visited the region. The Christian Science training of this religious trail blazer was almost unique in that part of California, and although she had originally planned to stay but six months, she decided to remain to minister at Riverside. Membership rapidly increased from an enrollment at Riverside of 10 persons in 1890 to 200 by 1901, when a church was erected costing $15,000; it was declared to be the community's most beautiful building.

Shortly after the introduction at Riverside, Christian Science came to other towns of southern California. The local press had to admit that its "growth in Southern California has been very rapid as it is less than ten years since the first teacher came to this section, and a strong and growing

[3] Mary Baker Eddy, *Science and Health, with Key to the Scriptures* (Boston, 1934), p. 425.

society is established in nearly every city in this half of the State."[4] Meanwhile, the *Christian Science Journal* in Boston published the testimonials of California health seekers as to cures. For the most part, these reports originated in Los Angeles, Pasadena, Riverside, Santa Monica, and San Bernardino.

In 1889 Angelenos were stirred by the case of the People vs. Reed, et al., involving the death of a child of the defendants which occurred during treatment by a Christian Science practitioner. Acquittal resulted, but a secondary outcome of the trial was the revelation that many prominent citizens favored the new sect. Their testimony under oath showed that most of them had been health seekers who found little benefit from climate and had turned to the pioneering religion. John D. Works, associate justice of the California State Supreme Court, testified that he was physically improved through Christian Science. L. M. Holt, the well-known southern California realtor, citrus expert, and journalist, said that his search for relief from lifelong dyspepsia had been rewarded through faith. Finally, Dr. A. Willis Paine explained why he had abandoned the practice of medicine when he discovered its limitations. Lawsuits benefited rather than checked Christian Science. The narrative of the Los Angeles case was used by the church to gain adherents. An equally famous trial with a similar outcome took place in San Bernardino.

The eleventh census of the United States substantiated all predictions of the faithful as to the denomination's growth. In 1890 there were 8,724 communicants or members, of which California had 814, or almost one tenth. Only Illinois and New York had more. In California the statistics were even more interesting. Only seven counties had any followers at all; three of these were in southern California, which contained the bulk of membership, 542, or two thirds of the total state figure. They were distributed as follows: Los Angeles County, 325; San Bernardino County, 150; and San Diego County, 67.[5] Of course, these figures were quite small, almost inconsequential nationally in

[4] San Bernardino *Daily Courier*, Nov. 17, 1889.

[5] H. K. Carroll, *Report on Statistics of Churches in the United States . . . 1890* (Washington, D.C., 1894), pp. 297-298, in U.S. Census Office, 11th Census, 1890, *Census Reports*, Vol. 9.

1890, but within a decade there were approximately 80,000 Christian Scientists in the United States, and the number had increased accordingly in southern California.

Indeed, Mary Baker Eddy was clearly aware of southern California. With the new century, she reorganized the church. In each state, branch churches were to nominate a branch committee on publications. There was one such branch for each state—except California. There, two committees were planned, one north of 36° N and the other for the important southern California region. An Eddy biographer, Edwin Franden Dakin, guessed that she had done this because California was so large geographically. An atlas, however, would have indicated to Mrs. Eddy that Texas was even larger and several other states had more people. The health quest is the almost certain reason for her action.

A permanent result of the migration of health seekers was the large-scale publication on faith cures. Southern California still has many such publications.

Other religious groups preferred to aid the health seekers with more conventional faith, hope, and charity. The greatest of these, charity, was stimulated, organized, and expanded in southern California where the proof of this devotional trinity was found in a generation and more of good works for invalids. The Roman Catholic Sisters of Charity had welcomed the poor sick to their little hospital since the late fifties. Yet the problem of the tuberculous migrant grew more acute, especially in Los Angeles. Some men saw a reasonable solution. One of these was a St. Louis industrialist, Nelson Olsen Nelson. For years southern California's health boards and medical societies had joined conscientious editors in calling for a philanthropist who might finance a colony for the area's pauper tuberculars. Here then was Nelson. He had made his fortune by building the largest plumbing supply company on earth. His interest in labor relations and profit sharing for both employees and consumers gave Nelson a reputation as labor's friend. In 1886 he had founded Leclaire, a model workingmen's colony near Edwardsville, Illinois. His town on the Mississippi was famous for having neither police nor paupers and the lowest death rate for miles.

In Los Angeles, Nelson heard of the Colorado Desert, publicized for decades as a health utopia by Dr. J. P. Widney and others. Nelson was convinced that its pure, dry air and sunshine would save hundreds of invalids stranded in urban areas. Accustomed to doing things on a large scale and succeeding in his endeavors, Nelson rapidly bought up 140 acres in the suburbs of Indio and laid out ten of them as a health camp. Here, thirty tents were set up, and two large mess tents appeared. Before long cows and chickens were supplying milk and eggs to sixty migrants. Invalids with a little money were supposed to pay $3.00 to $4.00 a week, while the destitute were charged nothing. Nevertheless, before many months had passed, the colony began to fall apart. Charity patients were being discouraged from coming, and only holdovers were treated gratis. Tents were sparsely furnished, and cooking was so uninspired that those who could possibly afford to do so ate in restaurants at Indio. At its best, Nelson's colony could not have unraveled the main problem; scores or hundreds of camps would have been necessary, and they would have needed a popularity that Nelson's oasis lacked.[6]

Clubs and welfare groups of all types helped carry on Nelson's insufficient samaritanism. The Odd Fellows were outstanding. In 1869 their San Diego chapter was chartered. Only four years later a local editor praised its work:

> We do not believe that any other Lodge in the jurisdiction has a better record for charitable deeds. Invalid members of the Order from all parts of the country came hither seeking health, and found here the brotherly kindness, the sympathy and active benevolence of the members of San Diego lodge. . . . Many came, alas! too late, to obtain relief from our health-restoring climate, and the last hours of the sick were cheered by the gentle ministrations of Odd Fellowship; the eyes of the dying brother were tenderly closed, and his mortal remains were carried to the place of burial in fraternal care. And the widow and orphan have blessed the order whose relief has been prompt and generous, and without ostentation.[7]

[6] Nelson's scheme failed because its haphazard organization provided insufficient medical care. Nelson's purse was strained. Interview with Dr. Francis M. Pottenger, Los Angeles, June 29, 1951.

[7] San Diego *Union*, Apr. 27, 1873.

Odd Fellowship officially came to Pasadena on December 30, 1885. After four years, that town's lodge could also boast of having cared for health-seeking brothers from 42 lodges and having spent $246.40 in the process. Its treasury had also paid nurses $180.50, while funeral expenses amounted to another $475.90. The Pasadena Young Men's Christian Association, inaugurated in 1886, was active through its Committee on Visitation of the Sick in nursing invalid youths, providing watches over their beds, supplying medicine, and in many cases securing medical personnel.[8] Like dozens of southern California towns, early Pasadena had its Charity Organization Society whose ladies individually helped the health seekers and as a group gave benefit entertainments to aid in financing their hospitalization.

Santa Barbara's Saint Cecilia Club was founded in 1891, its chief aim being to buy medicine and medical supplies and to rent shelter for the sick strangers congregated in the region. Dressed in white robes, seventy women brought to the sick "the soothing balm of shelter, rest, attention, and medical skill to alleviate their sufferings and restore them to health."[9]

Yet, such stopgap individual aid was pathetic if not actually defeatist. Genuine sanitariums were needed. The Cottage Hospital at Santa Barbara was an outgrowth of the critical situation and the attempts of civic-minded people to solve it. Optimistically one December day of 1891 the directors threw open the hospital doors. They were backed up with a capital consisting only of jellies, fruits, a few medical utensils, and several memorial rooms dedicated to charity in the names of relatives. Appropriately enough, Mrs. S. B. Brinkerhoff endowed one of these in memory of her physician husband, dean of Santa Barbara's boosters and a health-resort enthusiast. At the same time, on her deathbed a health seeker left $1,000 for the upkeep of another room.

San Diego could have used more hospitals. As late as 1890 the community of 30,000 had only two, the federal government's Marine Hospital and the Hospital of the Good Samaritan, the latter opened just the

[8] Pasadena *Daily Union*, Dec. 31, 1887; Pasadena *Star*, Dec. 28, 1889.

[9] Santa Barbara *Daily Independent*, Mar. 13, 1893.

year before with a quarter of its patients charity cases, mainly health seekers. Elizabeth A. Brewster was a heroine of the day. In 1875 she and her tubercular son had come to San Diego. On the same steamer were six other health seekers, who upon their arrival experienced great discomfort for want of quiet and suitable accommodations and care. Mrs. Brewster determined that when she had the chance she would do something to alleviate such suffering. Although her only son soon died of the disease, Elizabeth Brewster became convinced of the therapeutic value of San Diego's climate in the treatment of pulmonary troubles. While she lived, she personally aided health seekers and, at her death in the eighties, willed her Paradise Valley estate worth $20,000 to be used for the establishment of the Brewster Medical and Surgical Sanitarium, a memorial to her son. This nonsectarian institution thrived while newspaper accounts recorded its progress. By the end of the eighties the city had a free dispensary where the sick who received certificates from San Diego's churches or charities could get free treatment. More than half of these people suffered from lung ailments.

Hospitality was an old California custom. More than that, it was as vital as hospitals. A large number of sick migrants, poor young men who could get no work, were received into private homes where they were "as assiduously nursed as if they had been brothers or sons, until the end."[10] The quest for health was certainly not a purely masculine pursuit. Often women came west for their own sake or accompanied husbands or sick children. Most pitiful of all was the woman stranded when her invalid spouse died penniless in a strange community. Such persons could seldom get adequate work, and usually the only solution was the donation by individuals for a railroad ticket home. Churches continually aided these unfortunate strangers, but such kindnesses were on too small a scale. To alleviate the tragedy, some philanthropic women in Los Angeles decided in the spring of 1885 to organize the Flower Festival Society which would undertake annual floral shows, the profits of which were to be used to build a home for poor invalid women or destitute widows. As the shows immediately succeeded, a hotel was built where the im-

[10] Adams, p. 278.

poverished could live comfortably at very small expense. Some attempt was also made to locate jobs for these women. Near-by an orphan's lodge was established. Meanwhile, the annual floral festival developed into a big municipal affair, eagerly awaited by Angelenos and well reported by the press which yearly dedicated several columns to descriptions of the entertainment and decorations as well as to the organization's charitable accomplishments.

Los Angeles held no monopoly in flower festivals, although the idea for this sort of carnival may have originated there. In 1889 San Diego's own Flower Festival was first held, its proceeds going one third to the Women's Exchange, one third to the Women's Home, and the remainder for a day nursery. Pasadena had its Chrysanthemum Fair. Meanwhile in the latter community Charlotte Perkins Gilman tried to promote "a sort of residence shop where women out of work—there were ever so many stranded women there, who had come with tuberculous relatives—too late—could fill the frequent need of other tourists for sewing and mending."[11] She failed because of the lack of sufficient local support.

Nevertheless, organized charity usually did not fail. In 1887 a keen observer of such activities praised the conscientious workers, declaring that could all the facts be known in reference to their patient and gentle care of the sick, the whole country's gratitude would be awakened, since from every section of the nation people had gone to California for health.

At length the influx of health seekers would diminish, but organized charity had merely gained its first acceleration by this unusual migration. Charity, pioneering in a new country, had through a series of social emergencies proved itself "the greatest of these."

[11] *The Living of Charlotte Perkins Gilman: An Autobiography* (New York, 1935), p. 129.

CHAPTER VII

Mineral Waters

FROM BETHESDA'S pool through the holy wells of medieval Wales to Ponce de Leon's Bimini, history is filled with accounts of men who hoped to find health at medical springs. Although they were primitive in most activities, in the days of California's prehistory the Indians made effective but limited medical use of the local mineral waters. Knowledge of these spas became a heritage of the early Spanish settlers. After the United States acquired California, additional springs were rapidly discovered in northern California. Most famous were the "Geysers" of Napa Valley, which became the popular watering places of the fifties and sixties and remained well patronized by both the fashionable and the ill for several more decades. These springs were near the populated centers of that era. Because of scant population and poor transportation, southern California's watering places were to be neglected for a bit longer.

In 1862 Dr. F. W. Hatch tried to classify the medicinal waters then sufficiently known to warrant examination, visiting many of them throughout the state. Other authorities continued his pioneer study and presented their findings to the State Board of Health in an elaborate report of 1871. This attempt to find, analyze, and classify the springs of California was a highly technical task requiring very capable specialists. At the time such men were rare in the Far West. Yet, with such a stimulus this branch of medical geography brought forth the soundest research of the time to be instigated by the health quest. From the East and Europe came experts to test the springs. In their reports they often referred to the Bohemian Carlsbad, the waters of which were similar to those found in several California springs. This particular good fortune acted as an inducement for further study. One later-famous spring resembling Carlsbad in mineral content received the Old World spa's name, which is still preserved.

Since mineral waters had long been exploited in sections of the United States farther east for relief from rheumatism, nervous complaints, and skin diseases, the first information on California's watering places lured many afflicted persons to the West. Despite a background of data about older resorts, many sufferers came to tragedy. The pathetically egotistical often refused to consult capable doctors, preferring to depend upon their own diagnoses, which led them to bathe in or drink waters totally useless while their illness progressed unhindered.

Most injurious was another frequent error. Laymen believed that mineral waters were nearly uniform in their characteristics and, without medical advice, used springs sometimes with serious results. Here was another immediate reason for continuous research. Waters would have to be classified and their perils as well as their virtues determined. It would be a nearly endless work, since new springs were constantly being discovered. Fortunately, by the eighties industrious mineral analysts known as balneologists could safely list five distinct types of waters to be found on the Pacific Coast; these were: alkaline, saline, chalybeate, sulphurous, and thermal. These scientists compiled tables listing the percentages present of various constituents as well as water temperatures. Sometimes these data would be used by the unscrupulous for the making of unreasonable claims to cures.

The precise benefits of a given water were more difficult to ascertain than the mineral content. Even today the study of mineral springs in America has not advanced as far as in Europe. Alkaline waters, which we now know act as antacids, were used to alleviate intestinal ailments, catarrh, gout, and bladder trouble, but many other ailments were thought to be helped, too. Saline springs, which were sometimes thermal, were used by rheumatics and sufferers from scrofula. Today, they are noted for improving joint affections. Some observers of the period insisted rightly that chalybeate (iron-content) waters would aid in recovery from anemia but had no conclusive proof. Sulphurous springs were deemed helpful against lead poisoning.

Doctors were rightly skeptical of alleged miraculous cures by means of mineral waters of tuberculosis, heart trouble, syphilis, asthma, and even cancer. Such claims as these would only harm the sufferer as well as California's needed and deserved reputation. Alarmed at the growing number of tuberculars flocking to the springs for the climate and the heralded wonder waters, Dr. Hatch warned, "there is no fact in medicine, or in the history of consumption, more fully attested than that the treatment of that disease by mineral waters, especially the saline and alkaline waters, and their combinations, is positively hurtful."[1] Comprehensive chemical surveys awaited the new century.

For all the slowness of learned men to discover the true values of the state's mineral waters, the springs were the first health assets of southern California to be commercially exploited. The hotels built beside the spas became the first type of local sanitarium. By the sixties southern California was already developing such resorts. As the years passed, hotels and bathing facilities grew more elaborate, but even as late as 1900, when elsewhere in southern California the pioneering era had passed away and the frontier was gone, here the quality of lodgings still ranged from mere huts to palatial hotels filled with the most modern accessories and diversions for the blasé and fastidious invalids.

In the Santa Barbara-Ventura region, several important springs had been discovered early. In 1855 the Santa Barbara Hot Springs, enhanced by beautiful scenery, were found six miles from town by Wilbur Curtiss, appropriately enough an invalid. He secured the land, improved his own health by using the waters, and began to commercialize this natural resource. By the seventies a four-horse, daily stage from town

[1] Sixth CBH (1880), p. 31, in 24th Sess., II. Winslow Anderson, M.D., *Mineral Springs and Health Resorts of California* (San Francisco, 1890), is the best contemporary study of California's springs.

and good accommodations, including board and hot water baths for $3.00 a day, made his resort grow prosperous. At the age of seventy-two in 1878, Curtiss was still hale, appearing to be a fifty-year-old rather than a hard-working septuagenarian.

In the nineties the Santa Barbara Mineral Water Company developed the Bythinian Springs. Bathhouses covered several wells, and great lengths of pipe led to the main building where water was bottled for shipment to San Francisco and the East. The bottling works gave employment to local people. Veronica Mineral Water Company in the same region was also selling water. Dr. J. F. T. Jenkins, editor of the *Los Angeles Polyclinic,* finding the water beneficial asserted that it stimulated the mucous membrane of the stomach and bowels and testified, "I am forced to the conclusion that it has no equal of its class in this country, and will be favorably compared with the European, or other waters."[2] Los Angeles was an important market for Veronica Water.

About two miles from San Luis Obispo, Newcomb's White Sulphur Springs had by the eighties stimulated the building of a $10,000 hotel and several cottages, where sixty persons could be accommodated. Plunge baths and several other means of immersion were provided. The famous asthma and tuberculosis haven, Ojai Valley, boasted its own springs, with facilities somewhat more modest than those found at Newcomb's.

The most famous and best-patronized water cure in Los Angeles County was located two miles north of Norwalk. In 1874 while he was boring for an artesian well, Dr. J. E. Fulton discovered sulphur waters on his land. He immediately hired an analyst and at first allowed free access to a public which slowly but effectively became appreciative. Notable recoveries determined Fulton on a new course. Instead of letting strangers use his property as a type of public park, he would improve the estate, advertise it to the fullest, and turn it into a sixty-acre resort. Within a year he had built a two-story frame hotel with verandas on every side of both floors. It was almost always filled. At Fulton's and

[2] *Los Angeles Polyclinic,* I (1895), 133-134.

most of the better spas, one could find reading rooms, billiard halls, and other recreational facilities which patrons of the East's tourist and health resorts took for granted but which were quite an achievement of southern California's hotel building. Leisure wisely spent weighed almost as heavily as cure in the minds of those who took the waters. For example, Fulton had eight small, single-story bathhouses with front verandas. Each room had piped hot and cold sulphur water, and every compartment was supplied with a carpet, a cot for the weary invalid, and a chair. Fulton treated about 400 patients annually; most of these stayed a few weeks. With a well-balanced program of activity at Fulton's few were bored. By 1879 a windmill pumped water into a 50,000-gallon tank used as a swimming pool. By then Dr. Fulton was hearing the praises of recovered rheumatics and former sufferers of liver trouble and skin diseases.

By the eighties the founder was gone, but his namesake, Fulton Wells, had grown famous. Already the new management had provided free hack service from the Southern Pacific tracks to the springs, and the Santa Fe Railroad soon arrived near the spa, eventually baptized Santa Fe Springs. When the boom of the eighties burst, patronage declined, but the depression did not last long. As the new century began, so did plans for a new hotel and fifty modern cottages. The good fortune of proximity to two railroads gave the springs lasting popularity.

San Bernardino County's springs have received the boon of modern mass advertising. If one has heard of no other local waters, he has almost certainly heard of Arrowhead. Three generations ago the springs got their first publicity. Yet, long before the printed word arrived in California, Arrowhead Springs were well known; Indians used to bring their sick great distances to attempt to heal them in the waters. Then, in the 1850's a Latter-Day Saint colonizing group sent by Brigham Young to spread the Mormon word reached the site and settled San Bernardino. The little band was soon called back to defend Utah, but before its members left they had begun to call the natural formation on a near-by mountainside "the Lord's mighty arrowhead to punish the wicked." To latercomers, it seemed the Almighty's special symbol to bless the sick, and as the springs' fame for curing various cutaneous diseases and rheumatism grew, exploitation was inevitable. As early as 1858 a committee of the California State Agricultural Society reported:

> Some six miles from the town [San Bernardino] is a series of springs issuing from the side of the mountain, of such variety of temperature as to afford bathing to suit any class of nerves, and of such medicinal properties as to promise great benefit to the invalid-world who may be favored with its use. In a few years these springs will become as popular as the celebrated White Sulphur Springs of Virginia, and he who shall prudently invest there, making improvements as they are demanded, will insure a fortune.[3]

Before accommodations had been provided, invalids, not unlike their Indian predecessors, flocked to the spot, camping in flimsy tents and drinking the vaunted waters. San Bernardino's pioneer booster Arthur Kearney urged in 1874 the building of an invalids' hotel at the springs to house 150 guests. Such a place would profit local tradesmen and might benefit the country at large. By 1884 the resort, covering 160 acres, offered patients cold, vapor, steam, and mud baths as well as the pleasures of a 75- by 100-foot swimming pool. There were cottages for a few guests and unlimited space for campers. One of these patients, a staid Bostonian, described the typical treatments:

> I went into a room, undressed, took a sheet around me, and walked to the [mud] baths which looked like coffins on the floor, filled with mud. My bath had been fixed and tested with a thermometer. When they told me to get in—and Oh! how I dreaded that first step, it was just as soft and black as any mud you ever saw, only very hot—they packed me up and put my head [on] a pillow covered with a towel, so my hair was protected; then she daubed my cheeks and forehead, shut my mouth, packed the mud level from the chin, so you can imagine it was pretty deep on me . . . it was very heavy and the heat was very great; they gave me air and put water on my head. I staid in the mud ten minutes; they took me out. I was weak and faint; they gave me a

[3]*Transactions of the California State Agricultural Society . . . 1858*, p. 292, in *App.* Senate *Journals*, 10th Sess. (Sacramento, 1859).

> water bath. . . . I then tried a steam bath; they put me in a dry goods box over a natural spring. My head came through a hole in the top; my neck was bound with towels. I did not stay long, as I could stand the heat, but felt so clean after all was over. That Irish girl kept me drinking from every spring while rubbing me. When I got out of the mud bath I was blacker than any mulatto.[4]

Despite several costly fires which hindered the resort's advancement, the Arrowhead Hotel Company was incorporated, and during the boom of the eighties tourists found a four-story structure containing 100 fine single rooms and suites. Boredom was insured against by a large assembly hall for dances, lectures, and religious services, while a dining room offered good food, if not always the right diet for the sick. Even a regular post office and an electric power plant could be found there.

Other resorts of lesser fame enjoyed modest success near San Bernardino. For example, just east of Arrowhead, Harrison's ranch served a number of boarders who had come to try the waters on that estate. Farther along, "French Louis" offered his patrons in the Santa Ana Canyon rather poor lodgings typical of most outlying regions. In part, the mediocre housing conditions were compensated by the iron salts found in the canyon's springs. As roughing it was a necessity where log cabins were the only accommodations, few of the dangerously ill risked the hardships.

In the newly established Orange County, one might try San Juan Hot Springs near the famous mission for which the spa was named. Here again, facilities were poor even in the nineties. One had to live in a tent he had brought along for the occasion or carry "a gunny sack or rabbit skin for a bed," as nights were cool.[5] A local man with big dreams built a pavilion for those who might want to dance, preach, play cards, or dine formally. Most visitors, however, appreciated less pretentious ways and enjoyed the trout they had caught in near-by streams. A frequent guest described the scene at the local depot where one could while away his time viewing the constant arrival of health seekers:

[4] Pomona *Progress*, Dec. 2, 1886.

[5] Santa Ana *Standard*, June 14, 1890.

> an admirable study of human nature from the tall, lithe, sprightly, young lady, down through all the grades of deformity from diseases—stiff legs, swollen joints, contracted muscles, and most melancholy of all, those with tired out looks from overwork and study, looking aged prematurely—the incipient symptoms of softening of the brain was painful to behold, but in pleasing contrast to the above sorrowful looking picture of pilgrims to the springs. Some with old thick-set canes, hand worn from long use, straps, stays and rollers to assist in locomotion—all these after a sojourn of three to six weeks at the springs minus the old cane, rollers, splints, and crutches—our gaze was riveted on these for they are completely metamorphosed almost every one with a bundle or newly varnished canes carefully wrapped up but now he or she carries them under their arms and are able to carry any trunk or box to its destination—the depot.[6]

Not every observer could make such remarks, although mild benefits were general at San Juan. After a long period of obscurity because of poor transportation facilities, the hot springs were achieving their merited recognition.

Between South Riverside and Elsinore were the Temescal Warm Springs, frequently visited by Angelenos after the seventies. The tepid waters found there were strongly impregnated with sulphurated hydrogen and magnesia with some traces of borax and iron, but as late as 1887 there was still no exact analysis available, personal experience being the sole reason for the continuous influx of the sick and hopeful.

A few miles away, Elsinore Hot Sulphur Springs captured most of the fame and the cash enjoyed in the region. A colony had been established there in 1883, and during the great boom it grew suddenly from a few score to a resident population of 600. As a consequence of this increase, the California Central Railroad built a new station to serve what it called the "finest bath house on the Pacific Coast." Even well-experienced tourists who could be both fastidious and voluble acknowledged Elsinore's facilities as among the state's most extensive and suitable. Within a radius of less than a quarter mile of the center of town were scores of bubbling sulphur springs, characterized by the unmistakable

[6] Ibid., Aug. 16, 1890.

odor of rotten eggs. Rheumatics especially made long trips to use the Fountain Bath House and its plunge and sweat baths. Elsinore's scenic attractions have brought tourists and health seekers ever since.

Farther down the coast tourists sought out Carlsbad, almost identical in mineral content with the world-famous European spa. In the eighties the California Southern Railway built a station and shipped out the bottled water of the newly formed Carlsbad Land and Mineral Water Company. This resort was to survive through publicity, a good bathing beach, adequate transportation, and a magnificent setting. Ontario and Redlands people favored Carlsbad and patronized the four-story hotel built near the beach. Gradually, the resort became the nucleus for a town. On the other hand, some proposed resorts, founded with optimism, soon declined. Fairview, south of Los Angeles, was one of these. Its promoters believed the resort might rival Elsinore and Carlsbad, but time proved otherwise, and even the name Fairview is virtually unknown today.

A local firm with better luck was the Temecula Land and Water Company, which laid out Murrieta in 1886. Like most other spas, Temecula had good hunting, an added incentive which would bring tourists who could thus combine recovery with the pleasures of outdoor sport. Dr. Henry Worthington of Los Angeles, himself a health seeker, sent many of his tuberculous and rheumatic patients there even before a settlement had been formally founded, and in 1887 he reported that cures usually resulted from these prescription migrations. Still another physician, Dr. A. M. Lawrence, kept atmospheric temperature tables for the Murrieta area.

For centuries the springs at Warner's Ranch had been a gathering place for ailing Indians, and soon after the United States acquired California, San Diegans of a different culture but for a similar purpose trekked the hinterlands. These Anglo-Americans might also try Tia Juana Hot Springs (now Tijuana) just across the international border in Lower California, where even before 1890 fairly good housing was being provided and the resort was boasting the Pacific Coast's best sand baths.

These were the main mineral springs of southern California during the last third of the nineteenth century. There were many others, a good number of them charted and analyzed but obscure and virtually unvisited. At all these forgotten minor watering places accommodations, transportation facilities, and the number of patients were far inferior to those of the above-mentioned places. Northern California had had a strong head start in exploiting spas, and in this period of pioneering, hers were still superior both medically and socially.

Mere recitation of the names of favored resorts and narration of times and places yield little of real significance. Indeed, the overall importance of the springs in comparison to the search for health as a general movement was rather small. Sometimes towns did rise where the healthful waters gushed forth, but no great cities ever appeared. The spas' very smallness and comparative isolation from larger centers were definite attractions. Both serious health seekers and the more casual searchers for relaxation were protected from the noise and tensions of urban life, which as early as the boom of the eighties Los Angeles, San Diego, and other large towns were beginning to acquire. Furthermore, desperation was not a factor at the springs, as it was in the tragic case of tuberculars whose sole concern was survival. Often, pleasure was an important secondary goal for the springs' bathers when death appeared relatively remote and suffering at most was only a major discomfort. In almost every instance, transportation was the deciding factor in the popularity of a resort. Glamor rapidly tarnished if good roads and railroads did not soon arrive at a recommended resort. It was a remarkable spring indeed that brought campers over rough trails to its uncivilized and uncomfortable site.

If the habitués of the local spas never became the problem that the tuberculars always were, neither did they contribute so much to the economic, social, and political growth of southern California. Those who took the cure were predominantly transient, their stay but a matter of weeks. Furthermore, the patronization of mineral springs was usually seasonal. Thus this type of health resort and its clients were neither heartache nor social pain to the region.

CHAPTER VIII

Prescription: Agriculture

MUCH WISDOM has been voiced and published concerning the manifold benefits to man offered by farming. Through the centuries the agrarian life has received abundant praise for its spiritual values, and from the days of ancient Rome its development of one's physical side has been proclaimed. In southern California this trend was continued. In that region in the 1870's land was still cheap, soil rich, outdoor life virtually inevitable, and the region economically in need of settlers and producers. Meanwhile, the railroads were booming land sales to farmers and logically as well as inescapably connecting climate and agriculture to attract the early influx of health seekers and convince them that the wise would be both healthy and wealthy if they bought their lands. Many of the newcomers could foresee little opportunity in white-collar labor, for which the local economy supported few jobs and the sick population already offered superabundant personnel. Professions presented a dark aspect in a country where lawyers and doctors were dense in numbers and sometimes desperate in their finances.

Farming provided far greater rewards to the invalid than almost any career in the urban business world. Two very literate health seekers offered detailed and pertinent advice to the would-be farmer. Beatrice Harraden, herself a San Diego lemon grower, cautioned that small ranches conducted properly had the best chance for fair returns. Larger ones involved too much work and money for the physically unsound whose bank account was usually as limited as his muscular energy. Her collaborator, Dr. William A. Edwards, added that ranching could provide occupation, amusement, and recovery. This new method of farming need not be entirely masculine, for Beatrice Harraden spoke for herself and others she had observed in telling about the many light chores an invalid woman could do on a ranch. Another lady rancher advised, "I regard fruit culture a very healthful and paying occupation, especially for women who have children, boys in particular, growing up to assist them."[1] Such honest testimonials by those who sought no reward but their fellow man's welfare were bolstered by almost endless real-estate advertisements in newspapers and pamphlets throughout southern California and in several other states. "Good soil, healthy location" was a typical and hackneyed phrase for a generation, but it brought results. Although the realtors' propaganda produced its usual effects, nothing compares with actual accomplishments in sparking a movement, and in the case of the sick who turned to the area's chief livelihood, farming, there were many success stories to report. Numerous cured health seekers were glad to publicize their recovery and credit recuperation to climate and farming.

American honey production had never been undertaken on a large scale until the late nineteenth century. Then southern California led in making it a comparatively great activity. It was natural that the most individualistic branch of so self-reliant an occupation as agriculture would develop without benefit of much commercial advertising. The health-seeking rancher in southern California was a frontiersman, and beekeeping was the most rugged and probably the earliest rampart of this farm frontier. He went where the land was wild and empty, virtually by necessity seeking out hillsides and protected valleys where thrived numerous plants ideally suited to honey production. Here, free from coastal fogs, flourished the tubercular, and a little farther inland in the dry regions or peaceful uplands, bees and asthmatics did their best. Fruit farmer and stockman had not yet destroyed the natural vegetation which was the beeman's destined profit. When these more prosaic men

[1] Quoted in "Agriculture as an Occupation for Women in California," *Overland Monthly*, 2nd Ser., IX (1887), 656.

arrived, honey declined in price and production, and the apiarist ceased to be a highly significant builder of California.[2]

The golden age of California honey was inaugurated by a noninvalid, John S. Harbison, who came west in the gold rush. Noting the enormous variety of California plants, the suitably dry climate, and a myriad of pastures, he promptly denied the theory advanced by some early visitors that bees could not live in California, and before the fifties were half gone, he was shipping hives from the East. During the next decade Harbison founded San Diego's honey industry. The landscape was transformed, as soon it would be by the Australian eucalyptus. Before long, in the San Fernando Valley, around Simi, and throughout San Diego's back country tireless bee hunters were reporting their profits. In a handful of years the area was being called "the very Paradise of the honey-bee."[3]

As profits mounted, the increasing number of health seekers turning to the new occupation found the work exceedingly light, suited even to old people, its initial cost low, and the upkeep practically nil. There were no irrigation worries, either, for the beekeeper. As late as 1904, when prices had risen considerably, about $30.00 still bought three or four colonies of easily handled Italian bees and the apparatus needed to raise them.[4] Fortunately, all the average apiarist needed was much time and patience, and if there was anything the invalid had, it was time. Seeking survival in the West, he had learned to be patient. Many health seekers became intensely attached to their lonely yet idyllic lives among the bees. Here was freedom at last, freedom from stress, freedom from the East's winters and humdrum labor, but freedom most of all from illness.

[2] In 1874 it was noted that "bee men alone have done more in the settlement of the County of San Diego this year than all other classes, and more than had been done in the past five years." *Transactions of the California State Agricultural Society... 1874*, p. 342, in *App.* Senate and Assembly *Journals*, 21st Sess., I, (1875).

[3] Ninetta Eames, "Bee Culture in California," *Overland Monthly*, 2nd Ser., XVII (1891), 121.

[4] An expert estimated that for $1,400 an invalid could buy 50 hives, a honey house, food provisions for 15 months, a horse and wagon, and other supplies and materials to establish himself in beekeeping. *California as It Is*, p. 186.

J. M. Hambaugh was one of these men. He had suffered from annual recurrences of the grippe in Illinois where chronic "catarrhal trouble of the head" compounded the discomfort not only for Hambaugh but for his wife and daughter as well. Like many thousands of other health seekers, he made the difficult decision to leave everything and everyone he knew and move his family to California. In May 1895 the Hambaughs settled in Escondido where the ocean breeze was effective even that far inland. There they found their home state and Nebraska best represented by the growing numbers of health seekers in the community. In 1897 Hambaugh had thirty thriving colonies of bees plus abundant health for all the family. Using the apiarists' widely read trade journal, *Gleanings in Bee Culture*, he called upon his fellows to witness the blessings of "California, the land for the sick."[5] While he served as a herald of climate, Hambaugh became a leading honey producer. The bee inspection law of 1905 was considered a masterpiece by his colleagues, who said his efforts had won it.

Hambaugh's statements were being substantiated a decade later by a local beeman, A. J. Cook, who used his thirteen years of residence to assure readers of his very informative column on California honey production that the area cured tuberculosis and asthma. Another California beekeeper, while granting the healthfulness of the region, corrected the oversimplifications of Cook. This was not the most favored region on earth for consumptives, he argued. Coastal southern California was not universally recommended by physicians; thus he favored the interior, the farther inland the better. For support the columnist told of a tubercular who recovered far back in the mountains of Ventura County. If this was a valid rule, then it was especially fortunate that by the nineties beekeepers were being forced deeper into the interior by the eastward-advancing citrus frontier. Abundant advertisements reveal that inland heights were almost universally recognized as healthful.

By the century's last decade, enthusiasts recruited women for the apiarist's art. Unfortunately, this encouragement seems to have been largely

[5] J. M. Hambaugh, "California the Land for the Sick," *Gleanings in Bee Culture*, XXVI (1898), 48-49.

in vain, for an authority explained that most apiaries were located in mountainous sections, away from society, churches, and schools; these were lonesome places to live, especially for women. Consequently many beemen remained bachelors.

Isolated, self-reliant pioneers as were the apiarists, they were nevertheless varied in background. Some, and probably the most successful, were experienced in their calling. Hambaugh and Frank McNay were outstanding examples. The latter had come to southern California from Wisconsin where, before a physical breakdown checked his career, he had twenty-five apiaries. At Eagle Rock he successfully began anew. Others were what older residents termed "tenderfeet." Among these untrained newcomers was George F. Weeks, a journalist on the New York *World,* who was stricken with tuberculosis in 1875. As one of Charles Nordhoff's many optimistic readers, Weeks set forth for California and stayed for a time at a crude foothill sanitarium near San Bernardino. After several relapses there he fled to a bee "ranch" and became a "herder," slowly recovering in the hot desert air among sixty-odd hives. Thus even in the seventies beekeeping was becoming big business, a capitalistic enterprise which employed workmen and competed with the more individualistic hermits. Still, hired help was as often semi-invalid as were the employers.

These men achieved the zenith of southern California beekeeping. During their golden age, 1870-1900, even national periodicals publicized the free and lonely picturesqueness of this lucrative livelihood. When he was still editor of the Santa Barbara *Press,* Harrison Gray Otis wrote of the bee farms as he knew them.

> A bee ranch consists of the shebang of the keeper, two rooms, more or less, and his beehives flat on the ground near by. The bee range is unlimited, taking in the whole mountain side, and is free as air, for the busy bee can't be corraled, and is not amenable to trespass laws, fence laws, pound-masters or judges of the plain—thanks to the protection of his business end.[6]

Like others, he was amazed by the tremendous growth of beekeeping.

[6] Santa Barbara *Press,* Mar. 2, 1877.

By 1880 San Diego County produced the most honey of any county in southern California, perhaps more than any other in the United States, and probably a higher amount than in any other equal area on earth. In 1884 over a million pounds were shipped from San Diego harbor. That was an exceptionally good year; state-wide production reached 9,000,000 pounds. Yet in 1890 San Bernardino produced nearly 400,000 pounds, Los Angeles County 1,037,000 pounds, and Ventura County about 520,000 pounds.[7] In 1893 a promotional pamphlet estimated the annual southern California yield at 3,000 tons with over $250,000 as income from the area's 50,000 stands.[8]

Those who promised health also guaranteed wealth to the beekeeper. "Many will consider good health and promise of a competence as more than a compensation for the pleasure of city life," reasoned the Los Angeles *Times* in 1895.[9] By that date a number had grown wealthy. Yet in 1887 most beekeepers, still in the majority health seekers, got only four or five cents a pound for their honey. As white sage declined in acreage, the heyday of the honey boom faded. It had lasted about a generation, the health seekers' generation. Paradoxically, this activity which fled before more sedentary pursuits was the champion of the settlers. When beemen arrived, southern California was the domain of the stockman whose policy of open range discouraged farmers. While cattle ranchers would fight the settler who planned to plant crops, they would permit the quiet encroachment of the wandering apiarist whose charges ate no grass and required no fences. Beemen, however, were quite often voters, and they opposed the no-fence system. Their interests were more with the farmers and the development of transportation and a larger population, for, as with the farmers, these improvements were vital prerequisites to the advancement of their market.

Beekeeping still flourishes in southern California and is now completely allied with civilization, though its hopes will always thrive in the fields. It was still an undertaking favored by health seekers as late as

[7] Los Angeles *Times*, Jan. 1, 1895.

[8] Harry Ellington Brook, *The Land of Sunshine: Southern California* (Los Angeles, 1893), p. 38.

[9] Jan. 1, 1895.

1925 when Gene Stratton Porter, a resident of southern California, wrote *The Keeper of the Bees,* a novel telling of a sick man's turning to that way of life. Its initial stage, however, needs a sympathetic scholarly study to preserve a humble saga.

California as a center of citrus culture has long been established, in literature as well as on Wall Street. As early as 1870 local climate enthusiasts were calling the San Gabriel Valley and the area due east the "Great Orange Belt and Sanitarium." For example, the Reverend P.D. Young wrote, "By persons who have tried them all, it is thought that the climate of the 'Orange Belt' is better suited to weak lungs than that of the south of Europe or Florida."[10] Later a visitor, Henry C. Stiles, mused, "It is appropriate that in these countries where flock the sick, the infirm, the invalids of the world, for comfort and healing—should be grown this fruit which has such valuable medicinal qualities."[11]

As early as 1879 John Codman, an eastern tourist and even then well known nationally as an author of travel books, was impressed by the colorful fruit and the men who produced it. Experienced in observing human nature and picturing it with precision, he has left us an interesting vignette of the pioneer growers:

> most of them are a combination of ill-health, intellectuality, and comfortable circumstances. Orange culture is eminently adapted to their condition and circumstances. They can sit on the verandas of their pretty cottages—the refined essences of abstract existences—inhaling the pure air of the equal climate, reading novels or abstruse works of philosophy, according to their mental activity, from day to day, and waiting from year to year for their oranges to grow. Extremes meet. This is the sort of farming agreeable alike to literati and lazzaroni.[12]

This rosy sketch does not, however, mention struggling on a chilly winter morning with bonfires or smudge pots, but neither did the enthusiasts

[10] Santa Ana *Herald,* Dec. 31, 1881.

[11] National City *Record,* May 8, 1890. Although some health seekers became vineyardists, in this already long-established California activity they neither pioneered nor contributed as much as they did to citrus culture.

[12] John Codman, *The Round Trip* (New York, 1879), p. 56.

for this new type of farming. Citrus growing was virtually untried in America; similarly, the southern California climate was a novel experience. New, too, were those health seekers who engaged in orange culture, for most of them were inexperienced in this pursuit. They were lured west by advertisements and the letters of kind friends. An Ohioan, who had come during the great boom and now was writing east to advise a sick acquaintance, produced a missive as form-locked as if it had been stereotyped:

> You are overworked and run down by business confinement. Your wife and lovely children are being murdered by the cut-throat climate in which you live. Life and health are the chief riches, and first to be considered. Come out here where you can and will live out of doors one-half the time, the year round, and it will be the salvation of all of you. Take your $3000 and invest it in a little rancho within half a day's drive of town. You can get one already improved with comfortable buildings, large orange and other fruit trees, mature vines and plenty of room. . . . You will have no trouble in playing farmer. Take two or three years of this, and you will be again in the physical condition a man ought to have. Your mind will be clearer and your temper better. Your wife and children will be new creatures.[13]

"Playing farmer" became a most pleasant salvation for semi-invalids. Writing from ample experience in 1889, a grower observed:

> In the cultivation of this fruit (the most charming, healthful and lucrative of occupations, and decidedly the most fascinating) are to be found men from almost every walk of life. They have come from every State in the Union and from every Province of Canada, and not a few from the Old World. These men left the judge's bench and the banker's office, the editorial chair, the merchant's counter and the accountant's desk, the legal forum and the use of the lancet, the farm and the factory, the mechanic's tools and the professor's toga, for a mere congenial occupation and genial clime. They found both—and more. They found a larger measure of health both physical and mental; increased bodily strength with broadening of mind.[14]

[13] Los Angeles *Times*, Jan. 1, 1886.

[14] F. J. Gissing, "The Orange in Southern California," Ontario *Observer*, June 8, 1889.

A more detailed explanation of this "paradise for farmers" stated that:

> Those who had money bought at their pleasure improved property or land which they could convert into homes such as their taste and means dictated.
>
> But there were others who came here with very little ready money, broken down by disease or overwork, a family to provide and fresh from the office or workroom in the east. Ignorant of the first principle of agriculture, they were ambitious, hopeful and observant, ready to profit by the experience and advice of others, and our valleys are dotted with the happy and prosperous homes of those who have succeeded under such conditions as we specify.
>
> The merchant, lawyer, the doctor and mechanic; in fact nearly every trade and profession has its representative in the rank and file of California ranchers. Happy and contented in their work which has been congenial by experience, and return to health, they could not be induced to return to old sedentary life. . . . They look across the charming lawns and thriving fruit groves . . . and tell with pardonable pride the story of early trials and discouragements, when the battle seemed a failing one, with disease and poverty ranged against one feeble man's strength. But California climate proved the physician who brought healing, and Southern California soil the alchemist who turned dirt into gold.[15]

Once comparatively successful as an agriculturist and reasonably healthy, the invalid-turned-rancher was wont to reminisce, and being human, he tended to brag a little. His experiences were usually courageous and inspiring enough to deserve publication in the newspapers throughout the "Great Orange Belt and Sanitarium" where newcomers could take heart, and, incidentally, buy the farm lands then on sale. One interesting example, probably the best of them all, appeared in serial form as "An Invalid's Tale." The self-styled "Old Settler" advised:

[15]Banning *Herald*, Mar. 23, 1893. Jeanne C. Carr observed, "It was a singular fact that there was not a professional and hardly a practical horticulturist or farmer among them [Pasadena's founders]. . . . The worn-out physician found the fountain of youth in the pure California sunshine, which turned his grapes into delicious raisins." "Pasadena," in *An Illustrated History of Los Angeles County, California* (Chicago, 1889), p. 316.

> If the health seeker has no patch of land of his own, or town lot to work on, let him go down on his knees to the man who has, and beg and pray for the privilege of getting to work on it. Steady, persistent cultivation of the soil, in a pure atmosphere and under a genial sky, like we have here, will as surely save from destruction any lungs capable of salvation, as faith will save the soul.[16]

Although a "city bred man who never in his life owned a garden spot big enough to grow a daisy on, and moreover an invalid under medical sentence of death," in the early eighties he acquired thirty acres of virgin land in the Pomona region. Slowly and literally painfully he weeded half the area of brush. He later remembered, "My hands blistered, my feet got sore, my knees knocked together, often I had to stand fifteen or twenty minutes coughing and gasping for breath, every bone in my body ached, occasionally I caved in and went home." But he did clear ten acres. Simultaneously, his lungs began to feel stronger, so he turned to removing the stones, carrying the heaviest on his head. This, he stoutly claimed, strengthened the chest muscles! Thus continued the installments, and thus the farm grew and the man thrived. Eventually, he planted five acres of prune and pear trees, two patches of alfalfa, and other crops. Meanwhile he was building a farmhouse, barn, stable, and cow shed. All these agricultural and home-planning pursuits were realized in three years, but the medical ones took only eighteen months. By then the remarkable man's cough had vanished. Sadly he told of others who refused his work remedy and died.[17] More miracle than muscle must have saved this consumptive dynamo, since doctors of his day and of our own have found strenuous exercise an accelerator of tuberculosis.

Some scholars believe that Frederick Jackson Turner's "safety-valve theory" has been modified almost to the point of oblivion by other historians. Turner maintained that the Great West and its cheap or free lands had drained off the surplus labor population of the East's factory towns, thus preventing serious socio-economic ills. Later writers say that those pioneers who settled and farmed the West had not been city

[16] Ontario *Observer,* Jan. 26, 1889.
[17] Ibid., Feb. 9 and 23, 1889.

workers but rather experienced agriculturists farther east. This corollary may be generally valid, but for one area at least, and during about a generation, the health seekers of southern California established a splendid exception. True, they were not usually mill hands, but they surely had been urban dwellers, middle-class white-collar workers who knew nothing about agriculture. Yet, they were pioneers on a frontier, not a wilderness always, but in the early days an area still generally isolated and possessing a scant population and almost total political and economic dependence on eastern areas or San Francisco.

The invalids were strange "trail blazers," sickly, conservative, and sedentary for the most part; yet despite the absence of a rough and rowdy element, their agricultural rush was significant, and their courage outmatched that of some other frontiers. That so many of them did succeed, though amateurs and invalids, we must thank the good soil, a kindly climate, and the fact that even their healthy and experienced competitors in farming knew little about the citrus culture that most followed.

Successful farming by invalids was quite widespread. Its best and earliest example was Riverside. In 1870 that dry mesa area was a sheep and cattle grazing land, untouched by the white man. Even at 75 cents an acre the land seemed no bargain. That year, J. W. North saw the agricultural possibilities of the region, and the resultant Southern California Colonization Society determined upon founding a farming settlement. Theodore S. Van Dyke, southern California pioneer and health seeker, wrote of the attempt:

> Riverside and similar places were settled almost entirely by men of means who came here, first of all, for climate. Scarcely any kind of production was then profitable, beyond a mere living; but they came to stay anyhow, and having means enough to live on, they amused themselves with all kinds of experiments of all kinds and kept steadily in the belief that their work would one day pay. They are the ones who have worked out great problems of profitable fruit growing, changed all old systems of irrigation and cultivation and given us far better fruit and more of it with less water and less work.[18]

[18] *Citrograph*, Mar. 15, 1890.

That these well-to-do semi-invalids could afford to await the slow growth of citrus fruits instead of trying a fast money crop meant good profits in the long run. Yet it was not always a sublime occupation that they had chosen. Although everyone recognized early the healthfulness of Riverside, then considered an Eden for tuberculars, the area's agricultural destiny was not so evident. In 1872, when the first trees and vines were set out, the Los Angeles *Daily News*'s editor visited the colony and after investigating carefully and without prejudice wrote gloomily:

> The truth told, Riverside is not prosperous, and there can scarcely be said to be a promising future in store for it. Its settlers are principally invalids—sufferers from lung diseases—persons who have sought relief in its dry and semi-tropical climate, such having been recommended by eminent physicians as being peculiarly adapted for those afflicted with consumption.[19]

Furthermore, with most other people of the 1870's, he was convinced that tuberculosis was hereditary and thus Riverside would perpetuate a colony of invalids. Riverside's settlers were justly roused by the editorial, and their industrial association answered it, arguing that "though you spoke of our settlement as being composed of 'invalids,' we do not believe a settlement can be found in California where an equal number of people have performed more work in the same time. Let whoever doubts this come and see."[20] One citizen noted that:

> The health-giving powers of our climate are attested by many of us, more or less invalids, being capable of working in the open air almost every day of the year, and not feeling as we might feel in other places, drones in the world's great hive. In spite of prophecies to the contrary, we begin to feel that we are a success.[21]

Others observed that the large ranches were breaking up. In fact, the quest for health was emphasizing this small-farm trend, as semi-invalids

[19] Los Angeles *Daily News*, May 2, 1872.
[20] Ibid., May 9, 1872.
[21] Los Angeles *Herald*, Jan. 1, 1876.

often could not afford larger ones, and almost never could they cultivate over thirty acres. The ten-acre establishment was much more common.

Riverside's greatest fame by the eighties was as an asthmatic's haven. Walter Lindley upheld its benefits and humorously recounted that:

> We have a friend, a graduate of Harvard Medical, who went to Riverside badly afflicted with asthma. Believing he needed an out-door life, he went into the dairy business and circulated the following card:
>
> J. B. Johnson, M.D.,
> (Which Means Milk Distributor)
> Riverside, California[22]

Even Mrs. L. C. Tibbetts, who introduced the Washington navel orange to southern California, thereby guaranteeing Riverside's future, was an asthmatic. A San Diego *Union* reporter and health seeker interviewed her, observing that the visit became a mixture of their mutual asthma problems and orange growing, "And so we drifted, on and on, until I learned more about asthma than I ever knew about oranges before in my life."[23]

Riverside was certainly not unique in its being founded upon citrus and health; Redlands and Pomona were prominent among other localities with a similar foundation. Even before Pasadena was established, its site was noted for fertility and healthfulness. A tubercular originally from Boston owned a vineyard in the near-by foothills containing 100 small orange trees and 46,000 grapevines.

San Diego County gained many invalids who had suddenly but successfully transformed themselves into agriculturists. The back-country valleys, protected from the extremes of cold and wind, were most desired by these people. About the Guatay Valley a visitor commented: "There are many sheltered nooks where oranges grow. . . . Many are only intended for mountain homes for invalids. Two sisters who inherit [!] lung trouble have each forty acres adjoining and have built cozy

[22] "Riverside and San Bernardino: An Editorial Trip," *Southern California Practitioner,* I (1886), 377.

[23] *Citrograph,* Nov. 29, 1890.

homes and set out trees and flowers."[24] At Alpine there was considerable building by invalids on small tracts. Fine apricots grew there, in an area but recently considered marginal if even arable. Thus the health seekers were helping to open up, to farming, regions that would otherwise have remained untouched for more than a decade. The sick flocked to El Cajon Valley, and the land which had been cheap was immediately boomed. Some of these settlers hired workers; yet, as in the case of the "bee herders," a number of the farm superintendents were themselves seeking a cure. One such foreman thrived on easy outdoor work and labor direction, used the fruits he grew as basic medicine, and saved enough money to buy land which before long trebled in price.

Similarly, Escondido became a citrus center. Optimistic reports were legion. For instance, an invalid couple settled there in 1883, reclaimed the landscape from brush, quail, and rabbits, and by 1888 had transformed a hundred fenced acres into orchards and berry plots. Even Dr. Dio Lewis, sometime belittler of southern California's salubrity, noted the advances of Paradise Valley south of San Diego and the irrigation development and fertile land of one particular cured tubercular.

Pecuniary success was widespread. An "old Californian" returned to Kansas, developed lung trouble, and was forced to come back. On twenty acres near Azusa, "bare as an ordinary roadway," he set to farming. The extra expense of hired help did not prevent his making enough money to pay for the land, for which he was soon offered $200 an acre.[25] Since in the eighties many recovered tuberculars around Pasadena found employment spraying citrus trees, a wit exclaimed, "They came to cough and remained to spray."[26]

Edward E. Stowell, a 23-year-old tubercular editor of the Pomona *Progress*, observed shortly before his death:

> The invalid who purchased his small acre fruit farm a few years ago at a then seemingly high price, finds the same now not only increased in

[24] San Diego *Sun*, Aug. 15, 1890.

[25] Los Angeles *Herald*, Mar. 25, 1877.

[26] Carol Green Wilson, "A Business Pioneer in Southern California," *Quarterly*, Historical Society of Southern California, XXVI (1944), 153.

> value many times, but the products, as well, in such demand that the prices warrant a good interest on his labor and land, and he can continue the pleasant outdoor life that has given him renewed lease on the pleasures of earth.[27]

In 1889 another report on the "very liberal profits to be made in fruit growing" recognized the indebtedness to health seekers for the "phenomenal development of South California, emerging as it has, within the last decade, from a desert to its present condition of beauty and wealth."[28] To some degree this growth represented the false prosperity of the boom, although recent genuine progress in transportation and the development of new markets, added to the nascent citrus growers' co-operation in marketing, were about to assure more permanent and secure incomes.

Still, there were failures, financial as well as medical. Inexperience was primarily to blame for both. Invalids had heard impossible tales about enormous profits from little work in the fabulous citrus belt. They read of $1,000 returns per acre and thought that with a few hundred dollars they could collect easy rewards by planting any kind of trees with no real knowledge of climate, soil, or adaptability of citrus types. These men failed and packed home with their baggage bitter pictures of California's fruit culture. Other inexperienced health seekers did not realize that it takes five years for a budded orange tree to produce and ten for a seedling to begin to pay. A few were ignorant of the fact that mesa lands were best, nor did they understand southern California's basic water problem. Some, like a delicate Mesa Grande viticulturist, let aesthetic values decide their location, and although health came, dollars did not.

For those who had almost no money for land or agricultural implements, poultry raising offered less risk but a modest income. As one advocate of this livelihood wrote, "It is also a very handy business for an invalid with a small capital, who, perchance, must seek some light employment. The labor is light, if well planned; not more than two hours'

27 Pomona *Progress*, Sept. 22, 1887.

28 *Citrograph*, Nov. 9, 1889.

hard work need come in a day, and that day can come three days a week, and that with a flock of one hundred chicks."[29] At first the business was primarily for eggs. Not until Los Angeles pushed out its dense and growing population into a semisolid metropolitan area did chickens for meat become big business with San Fernando Valley its center.

By 1900 the first generation had done its work. Success outweighed failure, and numerous invalids had gained moderate publicity and good money through farming. J. J. Warner had come as early as 1831 in poor health. He prospered continuously until death ended his robust life in 1895. Henry H. Markham, governor of California, though unaccustomed to manual labor, copied his fellow semi-invalids in picking and drying fruit. Another health-seeking settler, C. H. Sessions, conducted during the nineties southern California's most scientific dairy.

The agricultural progress set by these early invalids still continues, but it is usually carried forward by the healthy, and its good results are insured by several state-sponsored experimental farms and laboratories. Big business and the power of wise investments are the tools of today's agriculturist as surely as the insecticide sprayer or latest-model smudge pot, and the total concentration is on profit. The invalid, by contrast, received none of this help and usually had but meager funds. Furthermore, he was forced to divide his efforts. He sought profits, of course, but his paramount aim was to regain health. Yet often he succeeded in both endeavors and thereby vindicated the booster's promise that here the wise could be healthy and wealthy.

[29] Pasadena *Chronicle*, Jan. 17, 1884.

CHAPTER IX

Those Who Came

DURING THE final years of the nineteenth century southern California gained considerable renown through the famous invalids who came to the region as visitors or settlers. All great movements have their giants. Some of them enter the procession already possessed of heroic proportions, while others gain importance through their role in some historic drama. This was true in the story of southern California as a health resort. For many years before the region was well known nationally, America's elite had gathered at Long Branch, Newport, and Saratoga for health as well as for pleasure. Comparatively early in the development of the Golden State, the Pacific Coast won its turn to play host to the mighty. The historian John S. Hittell, who would make his name on the Coast, joined the gold rush for his health, and California's most famous invalid guest, Robert Louis Stevenson, visited southern California briefly but preferred Monterey during his famous sojourn of 1879-1880.

Although the already famous were slow in coming, numerous men and women who arrived in the "New Italy" sick and unknown would soon help build the region politically, economically, and culturally. These health seekers were usually of the middle class and were comparatively well-educated people. Certainly physical condition had not made these people outstanding; yet their individual health was to be a very important factor, for in several notable instances the history of a cure directly and profoundly affected their adopted state. In all cases the recoveries of those who would later be leaders would help change the area indirectly. Grateful to southern California for real or supposed improvement in health, these usually highly literate people praised the region and encouraged other health seekers to come west. Meanwhile they themselves made original contributions to the section's typical and evolving culture.

Historians may have overlooked this unusual migration the basic aim of which was self-preservation, but health seekers did not forget history. James M. Guinn lost his health in the Civil War and then came to California by way of Panama. His interests were public education and American history. In 1881 Guinn was superintendent of the Los Angeles city schools; two years later he helped organize the Historical Society of Southern California. Although he would serve Clio, the Muse of history, and his adopted state by *A History of California and an Extended History of Its Southern Coast Counties* (Los Angeles, 1907) and *Historical and Biographical Record of Southern California* (Chicago, 1902), Guinn's more important contributions were made in historical journals, especially the *Publications* of the historical society of which he had been a founder. Fortunately, he preserved for posterity the records of many of the important historical events he experienced.

Even more interested in current history was Theodore S. Van Dyke, a midwesterner, who had come to San Diego in 1875 and sought recuperation by engaging in outdoor sports. He found what he hunted both in game and in health, as numerous writings soon made known. Spending several years in El Cajon Valley for recovery and becoming "the best informed upon its climate and resources, and the most disinterested of all residents," he supported real-estate promotion for the locality.[1] A result of similar advertising and speculation, the great boom was criticized and chronicled by Van Dyke, who wrote of it with honesty, accuracy, and a good deal of brilliant humor in his now classic *Millionaires of a Day*.

[1] Quoted in [Van Dyke], *The Advantages of the Colony of El Cajon, San Diego County, California* (San Diego, 1883), p. [i].

Charles Dwight Willard probably typifies all southern California's health seekers. Physically, his life was a series of many failures and a few brief periods of slow progress. Spiritually, it was the exact opposite. His old Chicago friend Dr. Norman Bridge wrote of Willard and the tuberculosis which altered his life:

> It changed his career; it transformed his life, and it transfigured his soul. Thereafter he lived the life of a soldier liable to go under fire at any moment. His fight against his infirmity was made with a courage undaunted by pain, and undisturbed by a thought of surrender.[2]

Willard's early letters show a whimsical, thoughtful nature, now practical to the verge of cynicism and again sentimental and wistfully sad. His moody side was tested in the winter of 1884-85 when he realized that he had tuberculosis. Willard had often talked of visiting his brother in southern California; now circumstances forced him to go. Settling in boomtime Los Angeles, he worked first as a *Times* and later as a *Herald* reporter, for journalism had long been his goal, even if California had not. Contact with civic affairs led him into an interest in refounding the Los Angeles Chamber of Commerce, a good cause which had been abandoned after a premature organization before the great boom. In 1888 he helped inaugurate the new chamber which would prove such a success. Willard served from 1891 to 1897 as its secretary, and his bust at the chamber's headquarters bore witness to his value in advertising southern California to the nation. With F. A. Pattee and others, Willard started the magazine *Land of Sunshine,* an incomparable booster of the region's climate, healthfulness, and history. This was a profound change of heart for an invalid whose earliest letters home revealed an unlimited disgust with the winter drizzles, the claims of unrealistic guidebook health-resort promoters, and an "overrated climate" in general. As the century ended, Willard was managing the Los Angeles *Express.* Besides his history of the local chamber of commerce, he wrote *The Herald's History of Los Angeles City,* a compact contribution to popular chronicles. In 1895 Willard was paramount in founding the Sunset Club with "60

[2] Bridge, *Fragments and Addresses* (Los Angeles, 1915), p. 176.

of the men with the most brains—not scholarship—to be found in this vicinity."[3] Dr. Norman Bridge and Charles F. Lummis were among those early members, and both were health seekers.

In 1895 Willard wrote of his friend Lummis, "condemned at one time to die of consumption and at another half paralyzed for four years, he has nevertheless made a record for himself in literature that so young a man can well be proud of."[4] Lummis made a fetish of exercise. His pedestrian feats are preserved in *A Tramp across the Continent* (New York, 1892), a record of his walk from Cincinnati. Los Angeles' famous librarian, a nature lover and conservationist, Charles F. Lummis in his journal *Land of Sunshine* exhibits his contributions to broadcasting southern California's health legend, while the Landmarks Club and the Southwest Museum are his monuments. There was hardly an aspect of the cultural life of southern California in the early days of this century in which he did not contribute something.

Lummis stressed many features of the bountiful region's actual or potential greatness, while George Ward Burton concentrated almost constantly on one, health. Well might he have, for, as Burton told the Los Angeles Merchants' and Manufacturers' Association in 1909:

> forty-three years ago in June I had undergone an examination by three physicians. . . . All three of these eminent physicians pronounced me a clear-gone case of pulmonary consumption, unable to weather two winters, if even one, in Wisconsin; that my only hope for half a dozen years . . . depended on my making quick progress toward California. I was examined the other day by the modern appliance of the X-rays, which is definite and specific, and not experimental. The revelation made by this exact science is that I now possess at this advanced stage of life not merely two lungs, but three, and all entirely sound.[5]

Burton the Three-Lunged was then a septuagenarian and still pounded out copy as one of the *Times*'s best writers. His newspaper articles

[3] Charles Dwight Willard to his father, Samuel Willard, [n. p.], Oct. 27, 1895, Willard papers.

[4] Ibid., Feb. 17, 1895.

[5] *Burton's Book on California*, pp. [v]-vi.

stressed the health factor to the extreme, while *Burton's Book*, a more elaborate literary endeavor, summed up his unbounded gratitude for and intimate knowledge of the region as a health resort. About 1914 I. W. Hellman, his old friend and a prominent California banker, told him, "I do not flatter you in saying that in my opinion your writings have attracted more population, capital, enterprise and industry to Southern California than those of any other newspaper writer who has ever penned a word in praise of that section."[6]

Such praise was unnecessary encouragement to one of Burton's optimism and dedication. Next appeared a lengthy poem, *Beloved California*, inspired by the health, climate, and scenery of "Mediterranean America." In presenting this book of verse, Burton again explained his purpose: "There are many invalids in the United States who might be benefited by the California climate as your speaker has been, and there are invalids being born and made every day in the year."[7] His closest approach to scholarship was *Men of Achievement in the Great Southwest: . . . A Story of the Pioneer Struggles during Early Days in Los Angeles and Southern California*, a biographical work published in 1904. There were greater health seekers than G. W. Burton, and many of them have been better remembered, but none expended more time and energy or appeared so tenacious a crusader for the encouragement of the migration of health seekers.

Semi-invalidism was common among southern California newspapermen of this period. One health seeker who would in time become most powerful of all local journalists was Harry Chandler. At eighteen Chandler was a promising student at Dartmouth College. Suddenly, his health broke, and he decided that his only hope for recovery would be a trip to Los Angeles. As it had been with others, the visit became a permanent residence for Harry Chandler, who at first worked on a ranch. After a relapse during a visit back home, he returned to Los Angeles and went to work for the Los Angeles *Times*. Chandler rose rapidly, married Harrison Gray Otis' daughter and upon Otis' death in 1917 became proprietor of southern California's greatest journal.

[6] Burton, *Beloved California* (Los Angeles, 1914), p. [27].

[7] Ibid.

Scholarly writing, too, was well served by health seekers as early as the eighties. In that growing health resort and incipient center of learning, Pasadena, Charles Frederick Holder became connected with the Throop Institute, founded by "Father" Amos G. Throop, an elderly scholar who had also come west as his own form of preventive medicine. Holder was professor of zoology at the little college, now the world-famous California Institute of Technology. Even in its earliest years solid scientific research was being pursued, and Holder was outstanding in bringing prestige to Throop Institute. He studied the islands and mountains of the region and in 1891 published his *Antiquities of Catalina,* which was followed sometime later by *An Isle of Summer: Santa Catalina,* a revision of the former work. A fellow scientist who had proclaimed southern California's healthfulness, Louis Agassiz, was the subject of a biography by Holder in 1893. His *All about Pasadena and Its Vicinity* (Boston, 1889) and *Recreations of a Sportsman on the Pacific Coast* (New York, 1910) also helped to broadcast the general theme which Holder always stressed—the wonders of southern California's geography. As it had Van Dyke, the life in the open which he knew so well suited Holder. As a result, it could be said that "probably no other scientific writer has contributed more to make known, both in America and abroad, this attractive portion of our great State."[8] It is not difficult to discover how Holder ruined his health by overwork, and at the same time benefited southern California, for in 1887 he was contributing a weekly letter to the San Francisco *Call,* New York *Observer, Post, Sun,* and *Times,* Hartford *Post,* Pawtucket *Gazette,* Worcester *Spy,* Lawrence *American,* Lynn *Transcript,* and Boston *Congregationalist* and *Herald.* A present-day syndicated columnist would have to be first rate to do better. His scholarly information, much of it readable and useful even today, complemented and bettered the usual florid propaganda of the place and period.

Among the most individualistic of health seekers was Abbot Kinney, former law student, medical student, merchant, botanist, cigarette manufacturer, and member of the United States Geological Survey. Both as a

[8]Newmark, p. 558.

tourist and a health seeker he had traveled through four continents and had settled in Florida. Yet no place seemed the ideal resort for this colorful globe-trotter. At San Francisco he had heard of the new colony at Sierra Madre. In spite of a well-earned skepticism, Kinney hastened south and established tentative residence in the new hillside community. Within four months he had become a permanent settler at his farm, dubbed "Kinneloa."

Kinney's several books differed widely in subject matter, but the most important themes dealt basically with outdoor life, health, natural history, and social problems related to health. All these tied in with the quest for health, which in his case had been fulfilled. Completely satisfied with his new home, Kinney wrote from Sierra Madre in 1882:

> Here the bright countenances of the people speak for themselves, and that is the best of all testimony. Now, that I come to think of it, I cannot see exactly why people should die in such a place as this. If there was any such thing as contenting mankind they would not. But I think that one ought naturally to live longer here and maintain vigor than anywhere else.

Speaking of the French Riviera, as almost everyone who discussed southern California did, Kinney explained:

> It surpasses us only in the additions man has made, such as beautifully planted promenades, drives and parks, for the invalid and stranger to wander in, and bands of music, theaters, casinos and reading rooms to amuse them. These we can add, and already the Sierra Madre villa, Santa Monica, Fulton Wells, and the hotels of Los Angeles have made a good commencement, and the work will go on.[9]

Kinney helped in large measure to bring better accommodations and cultural and entertainment facilities to southern California's resorts. Attempting to create in the "American Italy" something of the Mediterranean atmosphere, he founded Venice, counterpart of the canal city. The canals of California's Venice have long since disappeared, but the resort spirit of a sunny seaside town in part remains.

[9] *Pacific Rural Press*, Jan. 14, 1882, p. 18.

One man of science combined thought with action, thereby contributing to the development of southern California as only a few citizens had. This was Professor Thaddeus S. C. Lowe, of whom an acquaintance said, "He came here [Pasadena] to rest, but the exhilarating effect of the mountain air and tonic sea breezes recuperated his energies and his restless spirit launched into new enterprises."[10] Lowe had arrived in California in 1887 when he was already in middle life and had achieved nation-wide fame as an aeronaut before and during the Civil War. An attack of malaria in the sixties caused him to be almost constantly in pain for the rest of his life. Nevertheless, once in southern California, Lowe gained new vigor, and his energy was expressed in his numerous activities. He became president of the Pasadena Gas and Electric Company, vice-president of the Citizens Bank of Los Angeles, director and consulting engineer of the Los Angeles Lighting Company, director of the Columbia Savings Bank of Los Angeles and of the Citizens Ice Company of that city, owner of the Grand Opera House Block in Pasadena, and projector and president of the Pasadena and Mount Wilson Railway Company.

His connection with the last-mentioned institution was Lowe's most important work, the cause he loved most and which had the closest connections with health seeking. At a testimonial dinner given in his honor on August 23, 1893, fellow Pasadenans praised his railway project because it would make possible the transporting of invalids to the "health-giving climate of the mountain heights." Lowe had explained to them that he had been inspired to settle in the vicinity because of the various atmospheric changes available at different elevations. After several years in the near-by foothills, he knew the geography well. Meteorology had always been one of his studies, for it had been a prerequisite during his experiences in Civil War aeronautics. Always optimistic, in arguing for the proposed Mt. Wilson observatory he wrote:

> Under the auspices of such an institution, regular climatic observations from an elevation as high as Manitou in Colorado to the level of

[10]Bertha Knight Power, *William Henry Knight: California Pioneer* ([New York], 1932), p. 111.

> the sea, taken daily and given to the public and the medical fraternity throughout the United States, will draw a hundred fold more people to this coast than has ever been dreamed of by the most enthusiastic on the merits of our climate.[11]

Some distance away was the Lowe Astronomical Observatory. Tourists and invalids were served by the near-by Mountain House and Chalet.

The meteorological data gathered by men like Lowe were exploited by another institution, an instrument of boosting that outlived the classic boosters—Willard's chamber of commerce. It is historic justice that Willard's successor was another health seeker, Frank Wiggins, whom Willard called a hard worker and an intelligent friend. As a last resort, doctors had ordered Frank Wiggins to Los Angeles late in the 1880's. Once there, he was too weak to go about alone. Pessimistically, his physician gave him but a few weeks to live, but fortunately for Wiggins, his wife was a good nurse. With wise care and plenty of rest he slowly improved. As he did so, Wiggins came to understand the qualities of the climate upon which he depended. As a result of this he conceived the idea of advertising his "Southland." He became in time the dominating personality of the Los Angeles Chamber of Commerce, of which he was secretary for over twenty-five years. During that period it became the nation's largest chamber, and under his guidance it originated numerous methods of developing southern California's commercial potentialities. His was the idea behind the famous "California on Wheels" traveling exhibit which toured the nation displaying local products. America became conscious of the region's citrus and other agricultural products, its mineral wealth, and, above all, its remarkable climate. Wiggins was also in charge of every southern California display at American world's fairs from 1893 until his death. The poet John Steven McGroarty, who knew him well, said, "There is probably no more striking individual example of the possibilities of Southern California from the standpoint of health and human development than Mr. Wiggins."[12]

[11] *Inside Facts concerning the Pasadena and Mt. Wilson Railway Company* (Pasadena, 1897), pp. 27-28. Lowe is quoted at some length.

[12] John Steven McGroarty, ed. *History of Los Angeles County* (Chicago, 1923),

Local businessmen also had a large representation among those who had recuperated in southern California. Before the nineties wealthy merchants from east of the Rockies were already beginning to invest heavily in Los Angeles. Dr. John R. Haynes was one of these. In 1887 ill health forced him to abandon his home and profession. Although Haynes contributed occasionally to the *Southern California Practitioner* and other medical journals, he became famous in Los Angeles as a financier, connected with the Union Oil Company. The fortune which Haynes made through his oil interests was put to use in his crusade for good government. Contemporaneous with the Progressives and muckrakers nationally, Haynes almost alone fought for the recall principle in Los Angeles city and California state politics. Maintaining the tradition set by this outstanding health seeker, the Haynes fortune today aids sociological research in the form of the John Randolph Haynes and Dora Haynes Foundation. With headquarters in Los Angeles, the foundation publishes valuable monographs.

In the eighties Los Angeles began to assume the role of a commercial center in the Far West. Although the lack of statistics makes any but the most general conclusions impossible, it is probable that many men of means who permanently invested their eastern capital in the region from about 1880 to 1910 had come west primarily for their health or that of their family. We do, however, possess much more detail about several outstanding individuals who contributed heavily to southern California's economic development and who were also health seekers. Characteristic of a large percentage of these men was Andrew Mullen. A native of County Mayo, Ireland, Mullen had settled in Milwaukee until ill health brought him to Los Angeles in 1888. The bust which succeeded the boom did not long hinder his ambitions, for by 1890 the present firm of Mullen and Bluett, clothiers, had come into existence. In the mid-nineties J. Philip Erie came to Los Angeles for his health and while there built in 1897 the earliest automobile manufactured in the region. One need only scan the pages of the several "biographical histories" published

---

I, 354. Frank Wiggins' bust faced Willard's at the entrance to the Chamber of Commerce building. It was inscribed: "An institution is the lengthened shadow of a man."

in southern California from the late 1880's until about 1930 to find dozens of passages explaining how prominent merchants, manufacturers, and financiers originally decided because of illness to move west. Not all of these business leaders arrived in California wealthy, of course; some were virtually penniless and "grew up with the country" in the classic western tradition; others had made a good start farther east and continued their financial—and medical—progress on the Pacific Coast. It was more than poetic justice that the same climate which brought them to southern California would later give rise to trades and industries which to benefit from good weather would come west. Some of these same men, once more in good health, would invest in new undertakings which in their old home towns would have been either unusual or unknown.

Thus we see that practical men of business, seldom lured by a will-o'-the-wisp, were coming to southern California in about as great numbers as men and women of the arts and letters who might more logically be expected to follow a dream of health. Chicagoans successful in the world of commerce seem to have been especially susceptible to the publicity dedicated to the health legend. For example, on May 25, 1894, the Los Angeles *Journal* noted that a large representation of Chicago businessmen were coming to Los Angeles and its vicinity. They had begun to make heavy investments locally, while "their enterprise and enthusiasm have permeated every class of business and public improvement." Among these were S. B. Lewis, Holdridge O. Collins, Hans Jevne, W. T. Johnson, Earl B. Miller, and T. D. Stimson, all of whom would be remembered in the financial annals of southern California for their contributions to progress; every man on the list had come either primarily for reasons of health or at least to maintain the vigor he still possessed. Since it is impossible always to separate health seeker from climate seeker, since a single individual might come for several reasons, such as health, rest, pleasure, and business opportunities, it is probably safe to assume that even a larger proportion of business leaders were to some extent indirectly health seekers, too.

Perhaps as a counterbalance to the world of cashboxes and ledgers, the southern California quest for health stimulated the increase locally in the Republic of Letters—literature emphatically with a feminine flavor. Having suffered mental prostration in 1885, Charlotte Perkins Gilman went to Pasadena for the winter. "Feeble and hopeless I set forth," she later remembered, "armed with tonics and sedatives, to cross the continent." Among the orange groves of southern California she engaged in outdoor sports she had almost forgotten, made friends that she would always keep, and enjoyed to the fullest the casual yet ideally full life of an especially fortunate invalid, recovering "so fast, to outward appearance at least, that I was taken for a vigorous young girl."[13] Inwardly, she was active, too, for during the first year at Pasadena she produced thirty-three short articles and twenty-three poems. A literary product of Charlotte Gilman's misfortune was "The Yellow Wallpaper," a story concerning nervous breakdown. At least as well remembered were her poems glorifying southern California climate. "An Invitation from California" summarized the invalid's plight in a pertinent question:

> Aren't you tired of the doctoring and nursing,
> Of the "sickly winters" and the pocket pills—
> Tired of sorrowing and burying, and cursing,
> At Providence and undertakers' bills?[14]

Wherever Charlotte Gilman's work was popular, there the legend of health in southern California was carried.

Still another Pasadena writer, Margaret Collier Graham, went first to Anaheim in 1876 but soon moved permanently to the San Gabriel Valley. There her sick husband slowly recovered while he grew strawberries in the rich new farm land. The constantly increasing colony of invalids about her, and perhaps even her spouse, served Mrs. Graham as literary characters whose outlines she would ably sketch, their pathetic figures appearing in many of her better stories.[15]

San Diego attained some degree of literary publicity through the resi-

[13] *The Living of Charlotte Perkins Gilman*, pp. 92, 94.

[14] Gilman, *In This Our World* (Boston, 1898), p. 210.

[15] In *Do They Really Respect Us?* (San Francisco, 1912) Margaret Collier Graham said, "I have lived in California since 1876 and have in consequence no desire to go to heaven." P. v.

dence of Rose Hartwick Thorpe, author of the famous dramatic poem "Curfew Must Not Ring To-Night." Her marriage to Edmund C. Thorpe had been an exceedingly happy one until tuberculosis forced the couple to move to San Antonio for his health. Unfortunately, the Texas climate which agreed with the husband hurt the wife, so to escape the humid summer heat, they set out again, now for San Diego. There, on September 4, 1887, Mrs. Thorpe wrote to a San Antonio paper, "We feel confident that we shall regain our lost health here."[16] She praised the sunny port town and its potential greatness. The good times of the boom gave Thorpe steady work in his trade, carriage making. They lived for years at Pacific Beach where Thorpe was elected to the city council. Through numerous magazine articles, Rose Thorpe demonstrated her gratitude to San Diego. She joined the Ladies' Annex, a feminine auxiliary of the chamber of commerce, and supported its campaigns for civic improvements. Explaining her self-designed role at a meeting of the organization in 1890, she said, "The most enthusiastic among you does not love San Diego more than I, and while my work is almost exclusively sent to Eastern publications, I claim that I am doing San Diego a greater service than if I devoted my time to home periodicals." Comparing her products to those of the citrus growers who sent their best fruit east as southern California's most effective advertisements, she elaborated:

> into each story and essay and rhyme I drop hints of our magical sunshine, bits of description and carefully expressed convictions concerning this bright corner of the world. . . . I have only good words to say in reply [to fan letters], because San Diego has brought to me and mine the greatest good, the good of better health than ever experienced elsewhere.[17]

The happiness the Thorpes had found continued well into the twentieth century, and during those years Rose Thorpe faithfully boosted California.

Beatrice Harraden was a neighbor of Mrs. Thorpe in San Diego and

[16] Quoted in G. W. James, *Rose Hartwick Thorpe and the Story of "Curfew Must Not Ring To-Night"* (Pasadena, 1916), p. 24.

[17] San Diego *Sun*, Jan. 29, 1890.

a fellow writer, although her origin was Hampstead, England. Significantly, she wrote *Ships That Pass in the Night* (1893) about life in Petershof tuberculosis sanitarium of London. The book had given its author world-wide fame by the time she arrived in southern California seeking cure for an ailment similar to the one she had so well described in fiction. At San Diego she settled down to grow lemons and get well. Beatrice Harraden, however, did not give up her writing career. On the contrary, as it had other health-seeking authors, southern California served her as locale for new productions. Almost immediately after her arrival she wrote a novel, *Hilda Strafford,* which told of the desolate life of an English bride on a southern California lemon ranch. The story was not wholly autobiographical, for Beatrice Harraden remained a spinster. Later appeared *The Fowler,* another novel, this time set in Lower California, not far from her new home. The British celebrity created quite a stir in local circles. At Pasadena in 1895 a columnist described her as "plainly dressed and with somewhat of the languid air of both invalidism and the imaginative temperament that have made the author."[18] Yet she was practical enough to collaborate with Dr. William A. Edwards to write *Two Health-Seekers in Southern California,* a contribution to fellow sufferers in the form of detailed, down-to-earth advice.

Even at that time many were aware of the growing circle of minor literati in southern California. J. H. Gilmour of the San Francisco *Chronicle* exclaimed, "The number of artistic people living south of the Tehachapi is astonishing. . . . Southern California indeed invites to the pursuit of those arts which bring in their train mental healings and mental delights."[19] After the new century began and cultural institutions had increased considerably, the prolific naturalist and author, George Wharton James, built up an elaborate theory to explain the phenomenon. The splendid climate, wonderful scenery, romantic Spanish and Mexican historical background, and the promise of health were given as factors for the presence of so many gifted people. A modern writer who has ably delved into southern California literature gives a simpler version than

[18] Pasadena *Daily Evening Star,* July 12, 1895.

[19] San Francisco *Chronicle,* Mar. 20, 1892.

James's: southern California writers were "visitors, promoters, adventurers, and health seekers, the last category extending to some degree into the other three and certainly proving in the long run to be most productive."[20]

Drama shared honors with Hygeia in southern California. It is true that not so high a proportion of actors came to the region for health as did men and women of letters, but Helena Modjeska was certainly a star attraction among the few who did. In 1876 in Poland she heard about southern California from Henryk Sienkiewicz, who later became famous as the author of *Quo Vadis?* Mme. Modjeska was not in good health, and furthermore she was disgusted with Russian rule in her homeland and curious to see the New World. Doctors agreed with her husband that a sea voyage might restore her health, so the couple began planning a California colony modeled after Brook Farm. Southern California was all and more than they had hoped, but agriculture was not suited to these gentle Poles. Therefore, their estate, twenty-three miles from Santa Ana, became a mere pleasure resort and, at times, the Modjeska health retreat. In 1896 the tragedienne fled there suffering from acute thrombophlebitis. "Most of the physicians thought I would never recover," she wrote, "but thanks partly to my constitution and partly to the restful, balmy climate of our mountain home, I came back to my usual state of health."[21] At "Arden," a modest but well-proportioned home designed by Stanford White, she entertained fellow celebrities and made her house a tourist attraction until her death in 1909. She had fulfilled her often expressed wish to spend her last days in "the most delightful climate in Europe or America."[22]

Those "famous" health seekers who remained permanently in southern California were, as a whole, barely known outside the Golden State. This does not detract from their local significance. On the contrary, their lasting contributions helped make the region, if not themselves,

[20]Franklin Walker, *A Literary History of Southern California* (Berkeley, 1950), p. 106.

[21]*Memories and Impressions of Helena Modjeska: An Autobiography* (New York, 1910), p. 535.

[22]Pomona *Progress*, May 31, 1888.

world-famous. It is they who pioneered in letters and sciences, who organized chambers of commerce, who invested in nascent commerce and industry, and who in general led a society just emerging from the frontier status. On the other hand, southern California welcomed numerous "visiting invalids" whose fame and power were far from localized or provincial. They were great before they came. The scarcity of free time allotted to those rich in renown prevented their permanent settlement in a region still comparatively isolated and nationally insignificant. They could seldom stay long. Thus the shortness of these hurried trips for health often nullified any possible medical and psychological benefits to the individual health seeker, but brevity did not preclude the best advertising and glamorizing that southern California ever received as a health resort. That such famous personages would venture to the Far West proved to many easterners that the wilderness had indeed been tamed.

Most of these well-known pursuers of recovery arrived comparatively late in the period, about the mid-eighties or later, after the experiences of others whose roles were humbler had shown that the time spent might be worth while. Walter Q. Gresham was an exception. He was to be Cleveland's secretary of state, 1893-1895, but in 1874 was still a federal judge, worn out by hard work and plagued by a badly injured leg. As hundreds of doctors were doing with other patients, Gresham's physician recommended the newly discovered benefits of southern California. The judge spent the winter in Los Angeles, where he met his old schoolmate Horace Bell. Gresham was also suffering from dyspepsia, and Bell told the newspapers twenty years later that he had suggested ranch work as a cure. He set his old friend to work freighting manure to his orange groves, where for three weeks Gresham inhaled the "pure" rural air and daily harnessed a horse and drove his fertilizer wagon from sheep corral to citrus orchard—and meanwhile his stomach trouble improved. Later Gresham's widow wrote his biography, but naturally neglected this episode.

Disappointed in almost everything he had attempted during a long and momentous career, John C. Frémont in 1887 tried for the last achievement of all, survival, and in doing so made his final and least publicized trip to California. The explorer-soldier-politician's last years had been comparatively dreary, a weak constitution multiplying his troubles. As territorial governor of Arizona, 1878-1881, he suffered from the high altitude of Prescott and from mountain fever. Later, in New Jersey, Frémont was stricken with bronchitis.

Following his last trail to the West along a pair of iron rails, the Pathfinder arrived in Los Angeles on Christmas Eve, 1887. On that hope-filled holiday, he and his wife Jessie looked sentimentally "to Fort Hill, where forty years before our rescued invalid had planted a battery and raised our Flag. . . . To come back here was to renew younger life and find new strength." Slowly, the general improved, and his wife noted his cough was almost gone. An intimate friend of theirs confided later that on the night of their arrival he had thought Frémont looked doomed to early death. Mrs. Frémont wrote, "It had been a race for life against winter and Life won."[23]

The southern California press of 1888 and 1889 published frequent invitations extended by local towns desirous of having Frémont visit them, accept their proffered honors, and speak on various popular topics. Despite its preoccupation with the current real-estate promotion, southern California generously welcomed its reputed conquerer. Feeble health, however, prevented Frémont's acceptance of most of these demonstrations.

One function the aging hero could not deny. In the Armory Hall, January 21, 1888, a capacity gathering of 500 Angelenos inside and many more at the entrance welcomed Frémont. In his opening address the mayor of Los Angeles emphasized the migration for health, always an underlying current in local conversation. "What better commentary could there be on this glorious climate," he asked, "than the fact that the bold, dashing soldier, who forty-four years ago, so materially assisted in the conquest of our city, should return to spend the remaining days of a useful career?"[24] This was Frémont's seventy-fifth birthday, and when it came his turn to orate, he reminisced in the same spirit:

[23] Jessie Benton Frémont, *Far-West Sketches* (Boston, 1890), p. 22.

[24] Los Angeles *Tribune*, Jan. 22, 1888.

> I have never seen the birds go south when the winter came but I thought of California, and I never turned my thoughts to it but they were flooded with its sunshine. . . . Now, I have tried again. Just when, about the middle of the past December these winter storms broke over the East with a sort of blind fury, I struck my tent and started for Los Angeles. This time I will succeed unless the stars fight against me, or the rider on the pale horse cuts me down.[25]

The "stars" had always fought against Frémont, and although his health improved in Los Angeles, the "rider on the pale horse" was not far behind. In late 1889 business affairs called him to New York. There, just as he was about to receive a long-deserved pension from Congress, death came on July 13, 1890. Jessie Benton Frémont was far away in Los Angeles, there to remain quietly until her own end in 1902.

With the winter of 1887-1888 the Great Blizzard came to the East. That historically significant cold spell also resulted in the coming to southern California of more invalids than had ever arrived before in a single season. One of those who came was as famous as and decidedly more powerful than Frémont. He was the "Napoleon of Journalism," Joseph Pulitzer, who had been stricken blind and was immediately ordered by his doctors to stop work and go to California. Pulitzer soon found that his long trip had been in vain. Desperate for something quotable on the "health-giving climate," local newspapermen kept him constantly moving about. After only a day he had to flee Riverside and its persistent celebrity hunters. Yet, the "balm of America's Gilead," as his biographer called it, had not failed; rather, southern California was now in too close communication with New York to assure the journalist complete rest. Therefore, in April 1888 he left for New York. Pulitzer returned briefly a few years later, but meanwhile he had become America's most famous invalid, a veritable "career" health seeker. He traveled the globe, stopping only a few days or weeks in one spot and sometimes crossing the ocean and returning without a pause. Yet he never found the recovery he sought, and death ended his wandering in 1911.

[25] Ibid., Mar. 11, 1888.

A man to measure beside Pulitzer in journalism was Joseph Medill, a prominent Republican, a former mayor of Chicago, and the proprietor of the Chicago *Tribune*, which he had made a national force even before the Civil War. As his health was poor, in 1889 he purchased a large foothill estate at Altadena with a "superb view of sierra, valley, plain, city and sea, and surrounded with that rare and luxuriant growth of trees, plants and flowers which only South California soil and climate can produce."[26] Despite the best of surroundings and medical care, Medill remained a semi-invalid because of heart trouble. Meanwhile he indulged in his favorite health fad, the drinking of snow water. To Medill, the source of this liquid, a large reservoir of melted snow in the San Bernardino Mountains, was a true fountain of youth. He drank huge quantities to "wash the lime out of his system," for Medill was convinced that lime was a cause of old age and a hastener of death.[27] No European spa had cured the journalist's rheumatism, but he claimed that at last this mountain water had succeeded.

For the same purpose that today's hucksters seek out the famous and secure testimonials for some product, health-resort boosters persistently bothered prominent invalids of the eighties and nineties. Ben C. Truman, the railroad land advertiser, interviewed Medill, who allowed himself to be quoted in an 1892 guidebook. "Southern California," he said, "is by far the most delightful section that I have ever visited. Its winter climate is all that could be desired. It is essentially restful and exhilarating for aged persons [he was then 71] and others who have become worn by work."[28] On November 21 of that year his mansion burned. Medill planned to rebuild better than ever, but his invalid wife and daughter died in 1894, and he lost interest in the project. Upon returning to Chicago he sold his property in Altadena. Once again Medill was forced to seek another health resort, San Antonio, Texas, where he died in 1899. His sojourn in California had lasting results. During his last years one finds frequent publicity in his great newspaper given to southern California and its climate.

[26] Banning *Herald*, Apr. 21, 1892.
[27] Los Angeles *Daily Journal*, June 8, 1894.
[28] Banning *Herald*, Apr. 6, 1892.

Neighbor to Medill both in the Middle West and in southern California was Andrew McNally of Rand, McNally & Company. He was the first of several noted Chicagoans to settle in Los Angeles County for reasons of health, having come to Pasadena in 1880. Later he encouraged a good many of his neighbors to move to Altadena. After spending every winter at his ranch, McNally announced each spring that he felt fine. Evidently he diagnosed himself correctly, for he lived on until May 7, 1904, dying at Altadena.

Still another refugee from the "Windy City" was the "world's greatest butcher," Philip Danforth Armour. Although in his later years Armour spent the summers in Bad Nauheim, Germany, he could find no more satisfactory winter home than at Pasadena. Suffering from myocarditis, he stayed at his son's residence there; the latter died in Pasadena in January 1900, and his father returned to Chicago to live out two more years.

This often fatal error of spending only the winter months in southern California was made by another busy Chicagoan, Eugene Field. Early in December 1893 he left home, hoping for recovery in Los Angeles. There he suffered from the notoriously unheated boarding houses, but in spite of discomforts and poor health, his gallant spirit was always bubbling with interest and good humor. The poet met an old friend, Norman Bridge, and true to type played a new set of pranks on him. He also was invited over by another congenial acquaintance, Helena Modjeska. Upon returning home early in 1894, he wrote a fantastic article on his experiences in the Golden West. Field joked about mountain lions in the great actress' fields near Santa Ana and claimed that the real Ramona still survived in the prosaic reality of a train vendor. Unhelped by the California climate, the Poet of Childhood died in Chicago in 1895.

In the spring of 1892 Carter H. Harrison visited southern California. Like Medill he was a former mayor of Chicago and a newspaper publisher, but unlike him, a healthy man. At Riverside he praised the climate and urged the providing of "beautiful parks everywhere to add to the attractiveness of the land which is to be the sanitarium of the United States."[29] His publicity was effective back home, evidently even

[29] Ibid., Apr. 21, 1892.

upon his family, for in 1905 Carter H. Harrison, II, following his father in being a five-time mayor of Chicago, settled for four years in southern California for the health of his own son.

James Donald Cameron, United States senator from Pennsylvania, more commonly known as Don Cameron, and recognized as boss of his state, also came to Los Angeles for health in 1885. His visit helped advertise the region in the nation's capital.

A more important sojourn was that of the chief justice of the United States, Morrison Remick Waite, who came to southern California for recuperation in the eighties. Later, his sickly wife, Amelia Warner Waite, was persuaded to go west with her invalid sister. Her stay in Los Angeles was made pleasant by an uncle, the famous pioneer J. J. Warner. While she was away, the chief justice, too, was stricken in Washington, but rather than alarm his wife, he did not write her to return. More ill than he realized, Waite died in March 1888. His colleague on the Supreme Court bench, William Burnham Woods, served on the highest court from 1880 to 1887. He spent his final winter, 1886-87, in southern California but did not find the good health he sought and upon his return to the capital rapidly grew worse and died the following May.

Since the advent of American constitutional government, the families of presidents and former presidents have been the center of popular attention and, in some cases, idolization. Through the large-scale quest for health in southern California, two of these celebrated families were welcomed to the Pacific Coast. Just after the Mexican war, U. S. Grant had been stationed in California. He was later to be the first president to visit California, though not while chief executive. His son and namesake, Ulysses S. Grant, Jr., gained his most important fame in the region. In 1893 for his wife's health the younger Grant took the family to San Diego. The pilgrimage, like that of so many other households upset by illness, became something of a domestic mass movement. Josephine Grant's father, Senator Jerome B. Chaffee of Colorado, visited San Diego for the climate and to see his ailing daughter. Then in 1897 the late president's widow, Julia Dent Grant, settled among her family there. She, too, recovered her vigor while basking in the well-heralded sunshine of San Diego. Inspired to record her stay, she prepared a book illustrated by photographs by her son, Jesse. Ulysses, Jr., remained permanently in San Diego where he was reported at one time the county's wealthiest taxpayer. He was more successful in real-estate promotion than in the perilous field of politics, although California Republicans for a time thought of advancing him for high office. A memorial to his father, the U. S. Grant Hotel became one of San Diego's best attractions.

San Diego welcomed another presidential family, but in the case of the Tafts the situation was reversed; the parents, not the offspring, were the residents. In 1889 Alphonso Taft, former attorney general, secretary of war, and diplomat, contracted typhoid fever which gave rise to cardiac asthma, as the doctors called it. He migrated to southern California with his daughter, Fanny. Alphonso Taft's physician was William A. Edwards, Beatrice Harraden's collaborator. While treating his patient, Edwards fell in love with Fanny Taft, whom he soon married. Before Alphonso Taft died, Edwards told him that he believed his son William could easily get the presidency. He predicted correctly and later became the brother-in-law of a president. Had chronicles of that day been less reticent, we could probably list many more romances of both the humble and the great which save for the pursuit of health on the part of young people or their families would never have taken place. As for the Tafts, Alphonso's widow stayed on in San Diego for quite a long time and added her praise to the record of the health seekers' testimonials.

Without doubt the arrival of invalided notables snowballed interest in southern California. Comparative isolation, a more than mild provincialism, and an "inferiority complex" regarding the area's economic colonialism, a condition common to the West of that era, all prompted extensive advertising by the local press when climate was still the only important commodity. Logically, then, many citizens were aroused whenever rumors of visits by famous health seekers were circulated. Even if these proposed trips did not materialize, the prologue of favorable publicity was good for business.

The case of James G. Blaine's illness may be used as a classic example of this sort of publicity. Bad health, increasing political friction within his own party, and the satisfaction that his greatest works were completed led the secretary of state to resign in June 1892. Blaine of Maine had been the Republican presidential candidate in 1884, and many hoped that he might be again this year. Unfortunately, he was in his last illness. By autumn his family and doctors realized that he could not live long, although he rallied and talked of spending the winter in southern California. His political and personal friend, Joseph Medill, offered him his Altadena mansion as a retreat, and so Blaine's itinerary was arranged. When Medill's house burned, the plans to winter in Pasadena went on unchanged. Blaine would leave as soon as he could abandon the sickbed and travel by easy stages.

When the news reached southern California, papers from Santa Barbara to San Diego featured the story. The Banning *Herald* analyzed the situation, saying, "The coming of the famous patient to Southern California will turn to this part of the Union as much attention of the civilized world as have the visits of Bismarck to Emms, the vacation of Gladstone to Hawarden."[30] Young, energetic Redondo Beach, optimistic in its rising popularity, was jealous of the older resorts. Its press predicted:

> Mr. Blaine no longer enjoys robust health. But when a friend who spent ten months at Redondo Beach, last year, told him that this place affords a climate the year round which is superior to Bar Harbor at its best, the greatest of American statesmen at once decided to make Redondo his future home. So this seaside paradise, not Los Angeles, is to be the future Mecca of the Republican party.[31]

This was merely wishful thinking, of which the seaside resort held no monopoly. In December, Blaine's condition was reputed to be better. He would leave for Pasadena in January. Unhappily for southern California, as well as for Blaine, the statesman was dead in January.

California Assemblyman George W. Knox rightly observed that

[30] Dec. 8, 1892.

[31] Redondo Beach *Compass*, Sept. 3, 1892.

eastern governors, judges, millionaire merchants, men of letters and the sciences, artists, and important persons of a score of followings had come to southern California and upon returning home had almost invariably become what some listeners called "cranks on climate." "Their assertions carry great weight," Knox emphasized.[32]

As for those southern Californians who became regional celebrities during the period 1870-1900, it is almost more difficult to list the ones not directly in the area for their own or their family's health than it is to cite the health seekers who became famous. Yet, it will probably always be impossible to say precisely how great was the influence of "giants" in building southern California as a health resort. The total weight of unnumbered letters sent home by thousands of common men and women, the "ranks" of the migration, in a slower, quieter, less noted but more notable way helped to bring many more health seekers to the region. Nonetheless, the more measurable economic effect of famous visits and residences was significant, and the health factor certainly did bring rare talent to an area where it otherwise would have been delayed for several years. In many cases, too, it kept such welcome personages there who, were it not for reasons of health, would have had no pressing need to stay on the Pacific Coast. In comparison to its size and importance, this quest for health gained for the section as high a proportion of true human eminence as did the greater gold rush for all California.

[32] Los Angeles *Tribune,* May 5, 1887. So many prominent invalids had come to Riverside that when John Wanamaker passed through he had to deny that he was ill.

CHAPTER X

The Role of Government

IF THE Jeffersonian thesis that the least government is the best government is always valid, then for several years southern California's health resorts existed in a utopia. Today's observer, inclined to accept some role of government in advancing public health and safety, would see no paradise in the chaos of the early influx of health seekers. Beginning after the completion of the transcontinental railroad in 1869, the invalid invasion found health provisions in California guided by the same twin spirits of laissez faire and self-help which characterized the health quest itself. The term "makeshift" must have impressed itself upon any contemporary witnessing the results of the lack of organization both in private resort and governmental activity.

Most numerous of all and therefore of greatest significance were those resorts established for the care of tuberculars. Yet seldom were the "sanitariums" of the seventies sanitariums at all. Upon arrival in California most invalids registered at convenient hotels, there to remain until more permanent quarters could be chosen. A large number of these newcomers resided permanently in such lodgings, and hotel owners quickly learned to cater to their "weak-chested" patrons. Unfortunately, few establishments were comfortable by modern standards, and the tubercular was liable to find himself in a prisonlike room, virtually airless and boasting a northern exposure; such a situation was long exempt from even the most rudimentary of government sanitary inspections. Those invalids robust in initiative chose farming as their antidote to early tuberculosis. They needed and received little help from public health authorities.

The seaside resorts were more fortunate than the sanitariums, for as ocean swimming was still fairly new as a fad and even more recent among Americans as a prescription, beach resorts were less congested than other resorts, and the danger of infection was at a minimum. Furthermore, diseases which might be aided by salt-water bathing were usually only slightly communicable or noncontagious. Although the danger was small, medicinal benefits were also slight at the shore, and physicians complained for years that no rules were taught the public for the proper exploitation of sea water.

The mineral springs offered different problems. Scattered over much of southern as well as central California, these spas varied as greatly in their administration as they did in chemical content. At the most famous of them, entrepreneurs had by the seventies provided excellent accommodations and, occasionally, resident physicians. Usually the proprietor of a less important resort, however, offered only a crude shelter and his own inadequate brand of medical advice. Dr. Winslow Anderson, investigator of every significant spring in the state, lamented:

> A medical man is not allowed to tell one of these sovereigns that he must get up at 6 A.M. and drink the prescribed amount of mineral water and walk the necessary number of miles before breakfast; eat the regulation diet and strictly follow the *regime* best calculated to improve his disease; it would jar too much on his sensitive republican feelings. Yet this is just what American watering-places and sanitariums need. *It is the only thing our California mineral springs need* to make them as successful in this treatment of the many chronic diseases as the spas of Europe are.[1]

As late as 1892 another doctor warned that there was still not a place in the state where one could send a patient with any real assurance that he would receive proper care. Sanitation was good during the first six weeks of spring when the rush to the watering places began; after that the supervision of the springs became lax.

[1] Anderson, p. 15.

The greatest danger still remained the stupid advertising which convinced the gullible that any disease could be cured. Until the nineties few springs had been tested, although numerous amateur chemists had made sample examinations. Doctors still had insufficient knowledge by which to prescribe unreservedly that their patients make use of a particular resort. Thus there were two solutions needed for the double danger of professional ignorance regarding the springs and an overoptimism concerning their waters, perpetuated by advertisers and health seekers. The state government was the only agency capable of undertaking the task of supporting and administering a complete analysis of every important mineral spring in California. Obviously, the laws must force watering places to have resident physicians in constant attendance. First of all, however, the public would have to become accustomed to public health administration in general. As a change in public opinion would have to be effected to achieve this end, the goal of sanitarians seemed distant.

After the temporary epidemics caused by the improvidence, ignorance, and crowded living conditions of the forty-niners, California had settled down to a generation of good health, and not until 1870, a date coincident with the acceleration of the veritable health rush, was the State Board of Health founded. The act establishing the board required that:

> ...State Board of Health shall place themselves in communication with the Local Boards of Health, the hospitals, asylums, and public institutions throughout the State, and shall take cognizance of the interests of health and life among the citizens generally. They shall make sanitary investigations and inquiries respecting the causes of disease, especially of epidemics, the source of mortality, and the effects of localities, employments, conditions, and circumstances on the public health; and they shall gather such information in respect to these matters as they may deem proper for diffusion among the people.[2]

Theoretically, these duties would require trained personnel, and the results would be highly rewarding. In the beginning, neither was the

[2] Quoted in Second CBH (1871/73), p. 54, in 20th Sess., V.

case. In the first place, county boards of health were not immediately organized, and local boards were not widely established until the eighties. While the state legislators were working out the powers and duties to be assigned to the state board, public hospitals outside the Bay area and Sacramento were in embryo, and private health resorts did not have to report their activities to anyone. Thus, most invalids were beyond the very narrow pale of official regard. Even within this restricted jurisdiction, affairs did not go well. The local boards of health, not established by state law until 1876, at first failed to fulfill their obligation to report regularly. In 1877 the secretary of the state board complained:

> The present law requiring the registration of births, marriages, and deaths, has, I regret to say, proved utterly ineffective. If I may judge from the partial mortality reports received at this office from other sources, scarcely a single county has made a full and complete return of even this item—mortality—as required by the law. . . . Returns, such as they are, have, in most cases, been made, but the information afforded has been so manifestly incomplete as not to justify their tabulation at this time.[3]

The lack of well-trained personnel for the gathering of statistics and the unavoidable slowness of putting new governmental machinery into action partly account for the disorganization and the incompleteness of the work assigned to the local boards. In a few more years their reports became uniform and reliable, thereby making both the public and California officialdom realize how numerous were the health seekers and, far more important, how astounding their mortality rate had become.

As citizens grew more public health conscious, they heeded doctors who continually insisted that, since California had among the best mineral springs in the country, their full value should be known. This appeal to favorable advertisement was irresistible. Nevertheless, there were some die-hards, men not especially opposed to the analysis of mineral springs but fearful of the articles of a bill which not only would have provided for testing springs but also would have established prohibitions

[3] Fourth CBH (1876/77), p. 37, in 22nd Sess., III.

against the adulteration of liquids, foods, and drugs. The latter provisions would have jeopardized their vested interests. Happily, the legislation had numerous friends in the Senate; it passed both houses and was signed by Governor George Stoneman in March 1885. This act created the office of state analyst; W. B. Rising was the first man to fill the new post.

Before he began to work, Rising's plans were virtually destroyed. The law which had created his position neglected to appropriate funds to enable him to make analyses or hire assistants. He could not possibly perform his exacting duties without trained help which required substantial compensation. In 1890 the State Board of Health succeeded in getting $5,000 inserted into an appropriation bill so that it seemed Rising at last could employ assistants for a period of two years. Then the liberal provision was stricken out by the motion of a senator who said that it was a "useless expenditure of the people's money, and if they wanted their springs analyzed, or their food supply examined, to do it themselves."[4]

The reform movement had not failed. Businessmen were coming to realize that thousands of American visitors annually went to the mineral springs of Europe or to Arkansas, and California was consequently losing their trade by default. In the nineties a public health official observed the times he was asked for information about the springs and reported that "it is humiliating to have to reply that no official analyses of any of the springs have yet been made."[5] To make amends for the long delay, the State Board of Health published the best analysis available, Winslow Anderson's careful study.

The other solution to the problem of mineral springs, the employment of full-time physicians, was slower in its fulfillment. Public opinion accomplished a great deal. Without legislative action, by 1900 most of the springs had resident physicians capable of advising patients in the general problems of the water cure.

The white plague was also a long-lived puzzle to health authorities. In 1870 the belief still persisted that tuberculosis was usually hereditary.

[4] Eleventh CBH (1888/90), p. 45, in 29th Sess., VII.
[5] Ibid.

Many laymen as well as their doctors suspected that the disease might be communicable, but controlled and conclusive experiments had not as yet proved that it was. Legislation regarding plagues and quarantinable diseases was considerable, but these enactments completely ignored tuberculosis. No laws were passed in any way regulating the activities of its victims.

Contemporaneously, great scientists were experimenting in Europe and America. By 1882 Robert Koch, a German, announced his success in isolating and cultivating the tuberculosis bacillus. This proved that the disease was indeed communicable. A few years after Koch's epoch-making discovery, Georg Cornet's work supported the theory that infection was primarily through sputum. Tuberculosis was contagious, but not highly so, and even in those pioneering days, most authorities insisted that common-sense sanitation and special sanitariums would remove any danger of epidemics.

The reaction to the events in Europe was world-wide. Common man and great physician alike became tuberculosis conscious, and the people expressed their new interest through the establishment of public and private organizations. Southern California was to have a peculiar role in this drama for the control of tuberculosis, for there the health seekers had concentrated consumption to a probably greater degree than anywhere else in the Western Hemisphere. Now, everyone seemed aware of the mortality statistics which local, county, and state officials were finally compiling with trustworthy accuracy. These figures proved that tuberculosis was California's archkiller, as it was the nation's. In 1883-1884 the death rate of phthisis reached almost epidemic proportions. It represented 18.9 per cent of the state's mortality. Dr. Peter C. Remondino believed that the recent ravages of the grippe had lowered the public's stamina, thus permitting tuberculosis to gain considerable ground. He was correct, but to view only the year 1884 was to be exceedingly shortsighted. It was the out-of-state tuberculars who were dying, as had always been the case. Few healthy residents contracted the disease, and those comparatively rare victims were generally of the poorest section of society. Recently arrived Mexicans succumbed to tuberculosis due to overcrowding and the extremely poor nutrition and sanitation they could afford.

The first political reaction to tuberculosis had been the movement for a state sanitarium. Quite naturally it was the pauper who first got public aid. The scheme to build a public institution for these sufferers was conceived long before tuberculosis became alarming. As early as 1869 the Board of Supervisors of Los Angeles County petitioned the state legislature to aid them in caring for the indigent consumptives, since "the County of Los Angeles is now maintaining, and has maintained for many years past, an hospital . . . for the support and care of the indigent sick . . . , the genial climate of which is such as to attract the broken in health from all parts of the Pacific States."[6] That year a contributor to the San Francisco *Bulletin* advocated "some place where such persons could be sent with another chance, at least, for life . . . we believe a large number of invalids would flock to it."[7]

Throughout the seventies public opinion began slowly to favor a state-supported hospital. Despite its unfavorable climate for lung troubles, many immigrant invalids resided in the Bay region. As a result of this influx, the San Francisco City and County Hospital was overcrowded. Neighboring counties had sent their tuberculars to San Francisco, because of the more advanced medical facilities there, and two thirds of the hospital's inmates were from these outside areas. A system of county institutions would be too expensive and ineffective, most experienced men believed; therefore California doctors favored a state hospital to relieve the Bay area's burden. As there would always be enough patients to perform the lighter chores, the proposed sanitarium could be at least partially self-supporting. Building costs, too, would be small, for health requirements called for light and low structures on a pavilion plan and tents for those in the early stages of tuberculosis.

In southern California a state sanitarium was even more fervently desired. For several years the Los Angeles *Express* had campaigned for a

[6] "Petition of the Board of Supervisors for an Appropriation for the Support of the Non-Resident Indigent Sick of Los Angeles County," p. [3], in *App.* Senate and Assembly *Journals*, 18th Sess., III (Sacramento, 1870).

[7] *California Medical Gazette*, I (June 1869), 214.

state hospital and reported the county physician's claim that 43.4 per cent of the patients in the county hospital were nonresidents and most of them tubercular.[8]

In April 1879 John S. Hittell, the noted California author, gave impetus to the movement through an article in the *Pacific Medical and Surgical Journal*. Finally, upon hearing that a medical delegation from the Middle West was coming to select a proper site for a sanitarium of its own, the State Board of Health took matters into its slow-moving hands. In February 1880 it petitioned the legislature for the establishment of a state sanitarium. Two months later an enactment set up a committee of three from the health board to determine a suitable location, investigate probable cost, and evolve a scheme for construction and management of the proposed hospital.

The committee pioneered in the gathering of statistics, for as yet local health boards had accumulated but scant evidence upon areas particularly beneficial for tuberculars. Following a uniform system, committee members studied a given region's temperature, humidity, elevation, prevalence of fogs, wind direction and velocity, and the supply of pure water. Before it was through, the committee had visited places in Napa and Lake counties and had gone to southern California to investigate the Sierra Madre range in Los Angeles County and the Santa Barbara and San Diego regions.

Research had been painstaking but probably no more difficult for the committee than agreeing upon a single site for the proposed sanitarium. Since the transportation facilities of southern California at the beginning of the eighties were comparatively primitive, it was agreed to place the hospital farther north, where the bulk of the population was located. The committee's nod went to Atlas Peak in Napa County. This decision was disappointing to Los Angeles County, for in its report the committee had said that the Sierra Madre region had unexcelled climatic advantages. Perhaps in fairness to southern California, the investigators suggested that a second sanitarium be established at Sierra Madre, but to no avail. The Los Angeles *Express* complained that Sierra Madre ought

[8]July 8, 1877.

not to be considered inaccessible, for when the Santa Fe line should connect southern California with the East, as it actually did in another five years, the rush of tuberculars to Los Angeles County would find any sanitarium in central California even more inconvenient than the politicians now considered the southern section to be. Now San Francisco and Sacramento received the brunt of the sick paupers, but in a few years the trend would be reversed. Few health seekers would go the roundabout route to San Francisco. Therefore, the strongest argument in favor of Atlas Peak was also the most potent one against it.

Sierra Madre did not mourn alone the loss of a great opportunity. Despite the logic of the plan to win northern voters, the legislature refused to act. The greatest point of dissuasion after the financial one was an argument that would survive another thirty years. It was that this costly state sanitarium, if it ever were built, would soon be flooded with the nation's tuberculars, and California would be penalized rather than thanked for its wonderful climate and Christian charity by becoming the dumping ground for the sick of America.

The health board's investigation had not been in vain. Other reformers refused to surrender until stronger factors than political economy proved them wrong. By 1892 a state sanitarium had become a matter of public safety; the State Board of Health recommended that a "State lazaretto for incurable infectious diseases" be founded.[9] The board reasoned that the only maladies coming under this heading were leprosy and tuberculosis. Since there were no more than thirty or forty lepers in California, and all but two or three were Chinese, it did not seem urgent to establish a "pesthouse" for a few Orientals. Besides, Caucasian tuberculars of that era would hardly have consented to reside among members of the Mongolian race. Temporarily, at least, the problem of a tuberculosis sanitarium remained unsolved. Since many specialists still believed that every type of phthisis required a special climate, no one locality would be advantageous to a majority of California's tuberculars. In 1895, however, several members of the state board proposed to buy 160 acres near Salton for a desert hospital for tuberculars.

[9] Thirteenth CBH (1892/94), p. 7, in 31st Sess., VI.

With the new century, better results were obtained. Keeping pace with medical knowledge, the long-established county hospitals were now opening special tuberculosis wards. They began to erect cottages and pavilion annexes. By 1908 the new Southern California State Hospital's McGonigle cottages were "of one story, well lighted and ventilated, and have a large connecting porch, which is also connected with the nurses' cottage, where the patients can sit practically out of doors."[10] About this time the movement for county-supported colonies for the indigent sick began to eclipse the older crusade for a state sanitarium. Advocates of tuberculosis colonies wanted "health resorts" similar to those favored by exponents of farming and outdoor living. In such a colony patients would receive the best medical supervision and enjoy the ideal sanitation and nourishment impossible elsewhere. It would provide patients with all the benefits of desert life and ranching without the hazards and exertion of these pursuits. Some suggested that several counties unite, as they were doing to form consolidated school districts, to support tuberculosis colonies. The state legislature might subsidize these settlements and conduct others in connection with mental institutions. Within a decade the state insane asylums did carry out the colony plan, somewhat modified by time and aided by funds.

Because of its growing notoriety as a "dumping ground" for tuberculars, southern California had failed being considered by the federal government when it sought a site for a military sanitarium. Climatic and medical discoveries would soon cause the section to lose more of its reputation as a healer—to the great advantage of Arizona and New Mexico. The whole movement for a national sanitarium for the benefit of the American people was a disillusionment. Southern California had to be satisfied with the United States military post in San Diego, early established because of the region's healthfulness. In the eighties that encampment showed the highest rate of good health of any post in the nation; this record accounts in part for its becoming a general hospital for the Division of the Pacific, one of the few roles that the federal government played in California's health migration.

[10] Twentieth CBH (1906/08), p. 30, in 38th Sess., II. For a diagram of such a sanitarium see Sixth CBH (1880), p. 16, in 24th Sess., II.

The closest thing to a federally supported health resort that southern California realized was the National Soldiers' Home at Sawtelle. Although certainly no tuberculosis sanitarium, this institution was a retreat for disabled volunteer veterans, and the benefits of the local climate determined its site. The growing population of California had made it evident that some government hospital ought to be built to care for those veterans deserving attention. In December 1884 the Board of Managers of the National Soldiers' Home at Hartford, Connecticut, recommended that a branch be established on the West Coast.

As soon as this news reached southern California, various localities vied for the honor. None was more energetic than the Los Angeles Board of Trade, which wrote lengthily upon the county's healthful climate. General Nelson A. Miles meantime announced that statistics would be gathered regarding the "Climate and hygiene of Southern California as well as a list showing the percentage of mortality in that section of the country."[11] Within a few months a committee arrived to undertake the inspection ordered by Miles. Convinced of the salubrity of the Los Angeles area, the examiners gave tentative approval to the Sawtelle site. General W. W. Averell, assistant inspector-general of the National Soldiers' Home, said that Santa Monica and the Sawtelle region had advantages found nowhere else:

> Climate is not a thing to be measured by the thermometer or the barometer [as many health-resort advocates had done]. It has got to be felt, and so I may say the climate of Santa Monica is to all those other places as gold is to brass. . . . A large per centage of the 17,000 occupants of the National Homes in the East, who are afflicted, will necessarily, and very promptly, be transferred to the Home at Santa Monica. A good many of them are getting old and feeble, and their lives would be prolonged by coming here. So you may expect a good many of them.[12]

By 1892 the new home already had 741 veterans. Describing Sawtelle as "the healthiest location in the United States," the enthusiastic

[11] Los Angeles *Social World*, June 12, 1887.

[12] Los Angeles Board of Trade, *Annual Report, 1888* (Los Angeles, 1888), pp. 83-84.

Los Angeles *Times* claimed that most of the 10,000 old soldiers in the nation would like to retire there. The federal government was by then appropriating $250,000 annually for Sawtelle, and most of this money was spent in Los Angeles County. The $66,346 in individual pensions also remained in the vicinity.[13] Of all the "sanitariums" in southern California, the Soldiers' Home contributed more money over a longer period to the regional economy than did any other.

The adage that half a loaf is better than none does not apply to truth. Nevertheless, there are always those who make irrevocable judgments on circumstantial evidence. Founded upon a like basis of half-truths was southern California's "white scare," a psychological reaction which some termed "tuberculophobia" and others called "phthisiophobia." By whatever name the phenomenon was known, it critically altered the general public's treatment of tuberculars and produced an episode which was far from happy in the annals of the region's social history. Before the nineties residents of southern California had shown individual health seekers a great deal of hospitality, even if they were destitute and threatened to be a burden to the community. Unfortunately, by 1900 county hospitals were overcrowded with tuberculars, one death in six was due to phthisis, and the area was getting a bad reputation for the disease. The current scare was nation-wide, a natural outgrowth of discoveries about the nature of tuberculosis. Public interest in California was naturally greater than in most other parts of the country, although everywhere Americans kept themselves relatively well informed upon the nation's number one killer. Like most Europeans, the American people had always been accustomed to tuberculars, and although their numbers both real and relative grew as urban living developed, not until well-intentioned publicity appeared did the public shun these unfortunates. Then, and for the first time, the invalids' liberty, property, and right to happiness were controlled. Dr. V. Y. Bowditch, one of the world's greatest authorities on the disease, protested before the American Climatological Association in 1896 against the "sweeping statements made as to the contagiousness of consumption and the barbarism and brutality

[13] Feb. 18, 1892.

which the laity are thus led into showing in their treatment of the consumptive."[14] Some physicians, he lamented, broadcast the gossip that tuberculosis was as contagious as smallpox and that sanitariums were pestholes, a source of extreme danger to the whole community.

The fear of tuberculosis was more effective than a learned man's logic. In fact, a number of southern California physicians led the movement to restrict the activities of the tuberculous, all in the name of sanitation. The issue of private rights immediately arose, and Dr. Henry Gibbons, Jr., president of the state medical society, said:

> At the outset one feels repugnance at the thought of making public record of a physical condition and resultant mental distress made known to him in his professional capacity. . . . It is well enough to prevent promiscuous spitting, to educate the public to the understanding that tuberculosis is mildly contagious . . . and to adopt any reasonable measures to prevent its spread, but to require public record of all cases, incipient or advanced, will rather serve to invite concealment, and seems to some extent a reflection upon the capacity of the physician.[15]

"The question is," explained Dr. Charles W. Ingraham, "how far can the Government carry legal measures designed to control tuberculosis, and not infringe upon the natural rights of American citizenship."[16] Some doctors believed that tuberculosis, as it had been since 1893 in Michigan, should be legally declared a dangerous disease. In a strictly practical sense, however, these men felt that a registration law enacted before the public became better informed and accustomed to such restrictions would prove a failure, a dead letter unobserved by tacit agreement.

The chief expert on California's mineral springs, Dr. Winslow Anderson, favored the strictest regulation of all aspects of the tubercular's life, from cradle to deathbed, and even beyond. "Marriage of consump-

[14] "Sensational Statements regarding Consumption," *Journal*, American Medical Association, XXVII (1896), 107.

[15] *Transactions*, California Medical Association, N.S., XXVII (1897), 25-26.

[16] Ingraham, "Control of Tuberculosis from a Strictly Medico-Legal Standpoint," *Journal*, American Medical Association, XXVII (1896), 693.

tives should be prohibited," he warned. "Intermarriage of consumptives is one of the most baneful practices modern civilization countenances. Hundreds of deaths occur annually from this cause alone." He added that it was eugenically unsound. Anderson would have prevented tuberculars from becoming or remaining post-office or bank clerks or grocers. By no means would they be allowed to teach. He also advocated the compulsory cremation of tuberculous corpses. And while they lived, he urged, "Consumptives should always be isolated, and there should be established, under State control, hospitals and sanitaria for the segregation and isolation of the consumptive poor."[17] Anderson's totalitarianism would not only have prevented a tubercular from kissing; one would not even have dared shake hands.

Objecting to such severe restrictions as his colleagues were championing, one physician, W. B. Church, asked, "Shall every town maintain one or more pesthouses for consumptives, and pursue the same relentless course as is now taken with smallpox?"[18] This herding of tuberculars would probably aggravate the virulence of the disease and prejudice recovery. He predicted frankly that most radical methods would never amount to anything.

At least Dr. Anderson would have permitted out-of-state tuberculars to come to southern California. Other reformers opposed even this concession. In 1894 the president of the State Board of Health called for legislation to check the invalid migration, for, he warned, California was becoming the "tubercular sanitarium for the whole country."[19] Earlier, Dr. George M. Kober of Modoc County had expressed the stark view of those favoring a state quarantine:

> It is strange, but true, that in pointing out the danger to life, health, and wealth, incurred by the promiscuous mingling of consumptives with healthy persons, the Board [of health] should have been attacked by

[17] Thirteenth CBH (1892/94), pp. 323, 327, in 31st Sess., VI.

[18] W. B. Church, M.D., "Consumption," *California Medical Journal,* XX (1899), 308-309.

[19] Thirteenth CBH (1892/94), p. 241, in 31st Sess., VI. "The Fate of a Famous Health Resort" told how Menton, France, once full of happy people, was now a virtual plague spot. *Southern California Practitioner,* XI (1896), 69-70.

> heartless or thoughtless speculators, and even members of the profession, whose ignorance of well established conclusions is painfully apparent.

Using the simile of the stockbreeder who would not import diseased cattle no matter how well-bred they might be, he asked:

> Then, why should this glorious State be stocked with consumptives and their offsprings? Simply because we can sell a few town lots? . . . For instead of this State producing a people with mental and bodily vigor . . . we shall have a race weak in mind and body, and deeply tainted with a predisposition to consumption.[20]

His was an excellent example of contemporary opinion. Attempting to carry into reality Kober's stockbreeding analogy, a colleague requested the State Board of Health to consider the propriety of "quarantining against human beings and domestic animals with tuberculosis entering our State." He based his argument upon the theory that in certain localities in southern California the soil and air were "rich in the bacilli of tuberculosis," and that the disease outweighed the health-giving values of California while natives and older residents were permitted to become infected.[21]

Dr. Norman Bridge, who had made himself a nationally known authority on the disease that had struck him, assumed an attitude toward other victims of tuberculosis that was both merciful and exceedingly practical. With common sense he explained that the quarantine persistently proposed by some of his overzealous colleagues was ridiculous. Bridge asserted that it would require a guard of nearly 3,000 miles of border to keep tuberculars out of California, and four important railroad lines would have to be continually patrolled. A dozen or more harbors must be watched. Specifically:

> A medical inspector of skill above the average, and capable of making critical examinations of the chest with a stethoscope, and of the

[20] Eleventh CBH (1888/90), pp. 246-247, in 29th Sess., VII.

[21] Sixteenth CBH (1898/1900), p. 39, in 34th Sess., II.

> sputum with the microscope, would have to be on duty at each of these points. At the railroad entrances and the active harbors, two or three such officers would be required, also several deputy sheriffs or marshals, to execute the adverse orders of the medical officers. All passenger trains would have to be delayed several hours till all the passengers could be examined. The general appearance of the travellers could not be relied upon to tell who are dangerous consumptives, for some mortally sick ones look in the face very well, and nine-tenths of all tuberculous patients could easily hoodwink any inspector who should be less critical than I have indicated.[22]

If the contemplated law tried to keep out only the bad cases, medical officers would have to be clothed with broad discretionary powers. Large examination rooms and hospital wards would be needed at several points, Bridge thought. Since many trains entered California at night, all the passengers would have to be awakened and taken out under guard. In no sense would the purpose of the legislation be effectively achieved, for rigid inspection, costing the state between $30,000 and $60,000 annually, and perhaps as much as $100,000, would hardly be practical. If the examinations really proved thorough, and the tuberculars should cease coming, then the inspection would become useless, the law would soon be repealed by popular demand, and then once more the invalids would swarm in. In fine, the whole idea would create great unpopularity for California, its government, and the state's doctors.

Bridge did favor the state's transporting the indigents back to their homes the moment they became a public burden, but not before. A Los Angeles physician who thought as Bridge did based his plea primarily on humanitarianism, asking, "Is it possible that the great State of California is about to go on record as the most selfish spot on earth?" The results of such a decision would be this:

> Our sense of humanity would be held at a fearful discount should we even *seriously* consider the above resolution [of the State Board of Health]. Then, should we conclude with Quixotic ardor, to do the

[22] Bridge, "How Far Shall the State Restrict Individual Action of the Sick, Especially the Tuberculous?" *California and Western Medicine,* I (1903), 180.

> thing suggested, California would never again be the *one* bright spot and pride of the nation. . . . For once at least, the great railway interest, for selfish reasons, will be on the side of humanity, and object to a hold up at the State line of every incoming train.[23]

His strongest argument, a truth always stressed by the enemies of "tuberculophobia," was that the disease was only mildly infectious. Yet, on their own part, tuberculars did not organize to fight their well-intentioned persecutors, for, we must remember, they had little in common except their disease. They represented a cross section of the population of the nation, including all classes and regions, both sexes, and a variety of age groups. Needless to say, they did not crave publicity.

Yet, even without a lobby, the tuberculars were defended by far-sighted citizens and leading doctors. Dr. Francis M. Pottenger certainly hit southern California's conscience when he cuttingly asked if the proposed restrictions on "undesirables" would apply only to the pauper, or, as with other diseases, to all alike, including the wealthy tubercular who was once so welcome. A cynic might have concluded that it made little difference how empty one's chest was so long as his purse was full.

Disinfection of the quarters of those who had died of tuberculosis was a much more practical suggestion than most of those offered, certainly not unchristian and applying fairly to all classes. Yet, no law existed permitting health officers to do this. As the new century began, many informed observers attributed this "oversight" to authorities who were "too often subsidized by friendship, politics, or the more substantial, to allow them to interest themselves too conspicuously in sanitation to save lives when it might destroy votes."[24]

By 1900 researchers were coming at last to believe that victims of tuberculosis might be cured where they resided, granted the locality was reasonably healthful. Outdoor hospitals and sanitariums in the eastern states showed promising results. Consequently, southern California's physicians and health authorities no longer needed to feel that they were

[23] O. S. Lows, M.D., "Another View of the [Tuberculosis] Question," *California Medical Journal,* XX (1899), 309.

[24] Sixteenth CBH (1898/1900), p. 40, in 34th Sess., II.

acting ruthlessly if they discouraged pilgrimages to the traditional promised land of health. For instance, in 1904 an authority prophesied that eastern doctors would have to learn that "good use of a bad climate is much better than bad use of a good climate, no matter what the stage of the disease."[25] Five years later representatives of all the charities of Los Angeles County voted to cease aiding tuberculars from outside their county. They felt that had this step been taken years before, the medical profession east of the Mississippi would long since have stopped sending indigent patients west.

Soon after this measure was taken, the railroads were co-operating with the officials of Los Angeles in sending back tuberculars without means. Provided with baskets of food, these invalids could now be seen boarding eastbound tourist cars. They did not always have far to travel, for, according to Dr. D. C. Barber of the Los Angeles County Hospital, "Adjacent county officials think it a joke to present their indigent consumptives with a lunch and a railroad ticket and drop them upon the lap of our Associated Charities."[26] Hundreds of other health seekers, given a small purse for traveling by local charities or well-meaning clubs or churches in the East or Middle West, had been told to say that they were residents of Los Angeles County of more than a year's standing and therefore legally entitled to shelter.

In the 1900's officials were determined to solve the problem before any disaster might result. For a while they were unsuccessful, and their victory was never complete. The significant factor is that southern California, which at first welcomed tuberculars, hoping that even the pauper element would be cured and might become a stable, self-supporting group, now would accept only those who had jobs or capital.

While the health rush had slowed down, the era of effective sanitation began. The most controversial suggestion produced by the white scare was compulsory registration of all tuberculosis cases. When the reform was put into effect, it caused some suffering. Conservatives favored only

[25] George H. Kress, M.D., "Tuberculosis," *Los Angeles Medical Journal,* I (1904), 315-316.

[26] California Association for the Study and Prevention of Tuberculosis, *Bulletin,* I (1909), 30-32.

enforced notification of local health boards and disinfection of the premises in all fatal cases, but the state legislature enacted a law requiring the notification of health boards of all cases. A number of doctors and health officials disapproved of this comprehensive coverage, for it made public the required data. As a result, many hotels immediately refused to shelter invalids, for they disliked the resultant notoriety and the probable fumigation.

Governor J. N. Gillett saw the wisdom of notifying health officers, but he was against the proposed ruling regarding disinfection within certain time limits and under specific conditions of a room in which a tubercular had died. He sensibly noted that:

> The loss of business and the inconvenience and loss through disinfection, would cause the closing of all doors against consumptives. . . . They would become outcasts, and shelter would be denied them. Not only would they be excluded from all dwellings, but no person could travel with the assurance of receiving shelter unless he could produce a physician's certificate that he was free from tuberculosis. No landlord would rent apartments without the preliminary requirement of a clean bill of health from his prospective tenant.[27]

In any state but California such regulations would not have greatly alarmed landlords, but since health seekers were so numerous, vigilance became almost universal. Charles Dwight Willard knew firsthand the results of these ordinances. By 1911 when restrictions were fixed and prejudice was almost traditional, he described his feelings upon being

> turned out of house and home in a country where a consumptive is treated like a leper with every hotel, boarding house and house for rent closed against him. Even the "haunted" place where the Dwights [not relations of his] killed themselves, unable to get a tenant in 2½ years, was refused to us—we were so desperate we would even have taken that place of horror.[28]

[27] *Transactions*, Commonwealth Club of California, V (1910), 269.

[28] Charles Dwight Willard to his sister, Sarah Hiestand, [n.p.], Jan. 3, 1911, Willard papers.

His old friend Norman Bridge gave Willard and his family a haven, but others were less fortunate. In general, probably the unwanted pauper was enviable, for he fared better than did the middle class. As Willard noted, the really good sanitariums were for the poor, although these were always overcrowded.

If living quarters were difficult for the invalid to find, so also was disinfection difficult for the officials to enforce on a large scale. As late as 1909 sanitary precautions against tuberculosis were not strictly observed. The State Board of Health began making inspections of summer and health resorts, but in 1906 it had to report that the various resorts were so numerous and scattered that examination could be undertaken only of those involving complaints. Fortunately, the authorities could announce the willing co-operation of most resort managers.

Counties were usually less cautious in establishing restrictions than the state had been. For example, in the 1900's San Bernardino's board of supervisors forbade the transfer of the noted "Settlement" at Redlands to Mentone. As the new property contained an open ditch through which the municipal water supply flowed, the supervisors hurriedly passed an ordinance forbidding the operation of any institution for contagious diseases within 600 feet of an open water supply.

Following an illogical but very human trait, Los Angeles moved in several directions at the same time, not all of them beneficial. Fear had motivated action. In 1890 Dr. Frank D. Bullard, who as an interne in the Los Angeles County Hospital had kept a careful record of the incidence of phthisis, said that one seventh of the patients had tuberculosis; 47 per cent of these died, as they were in the last stages, and, he presumed, the proportion of this to other diseases was nearly as high outside the hospital as within, although the death rate was lower, since many ambulatory tuberculars returned to the East to die.[29]

The next year another physician of Los Angeles announced, "I presume it is fair to state that more than three-fourths of the population of Southern California are victims of consumption or have a taint of tuber-

[29] Frank D. Bullard, M.D., "Climatology and Diseases of Southern California," *Southern California Practitioner*, V (1890), 206-208.

culosis."[30] This calculation seems ridiculous. Like many others, he looked upon Los Angeles' crowded and unsupervised boarding houses as pestholes. He had known men who lived in such surroundings among the tuberculars and after a year or so lost their own health and had "all the aspects of tuberculosis." Perhaps poor ventilation alone would have caused this deterioration. Other writers, too, insisted that tuberculars congregated in the cheapest lodgings in Los Angeles, for they were often poor and possessed blind faith in climate, not domestic environment.

Such alarming statistics and half-truths caused the city council to act in order to guard public health. Early in 1904 that body made it unlawful to build any hospital where tuberculars were to be treated within 250 feet of a residence, school, or church. Vainly an attorney argued that "God Almighty gave us this climate, that attracts consumptives from all over the world, and we cannot keep them away if we would!"[31] In May of that year a less reasonable law was passed whereby all hospitals were forbidden to retain tuberculous patients, excepting only the county hospital and the Barlow Sanatorium. This move was the result of fear on the part of homeowners living near the Kaspare Cohn Hospital which admitted such patients. This sort of procedure was inconsistent with common sense, for institutions where hygienic rules were observed ought to have been encouraged. At the same time dangerous lodging houses and cheap hotels still openly violated sanitary requirements.

The railroads, for two generations attacked as villains by California's political and economic reformers, were pounced upon by the sanitarians, too. In 1896 the latter insisted that regulations must not only cover residences but also all means of transportation. Tuberculars should travel in special railroad cars which could be fumigated, and their laundry ought not to be mingled with that of ordinary travelers. In 1900 California still had no over-all code to regulate railroad sanitation, although since the early eighties the Southern Pacific Company had been accustomed to co-operate with state health authorities by inspecting its

[30] William H. Dukeman, M.D., "On the Subjection of Consumption," *Pacific Medical Journal,* XXXIV (1891), 23.

[31] Los Angeles *Times,* Jan. 31, 1904.

trains during the frequent smallpox epidemics. Since the nineties the Pullman Company had maintained a superintendent of sanitation. With public opinion focused on them, the railroads continued this far-sighted policy, and no serious complaint about their antituberculosis sanitation was heard.

As sputum is the chief conveyor of the tuberculosis bacillus, anti-spitting laws were inevitable. By 1898 Los Angeles and Pasadena forbade expectoration on the public streets, a significant move in two of southern California's chief health resorts. Even in San Francisco a similar ordinance was adopted because of the campaign of influential doctors. In 1907 a state antiexpectoration law came into being. Not only sidewalks but such dangerously infective locations as ships, cars, and public buildings were included.

The best discipline is voluntary and the finest law self-imposed. Public health authorities realized that health seekers would naturally resent restrictions but felt that they could be won by the common sense of simple and clear instruction regarding the character and dangers of tuberculosis. After the nineties, the decade of the white scare, the State Board of Health, aided by several charity organizations, began a campaign to educate tuberculars as well as the general public. The victory of these people vanquished the terror of tuberculosis, and invalids began to observe hygienic precautions. Operators of jerry-built sanitariums came to see their doom. After 1900 there departed those "conscienceless, rapacious, predatory rascals who fatten upon the credulity of suffering humanity, and who throng . . . as thickly as the gray wolves of the plains upon the heels of a wounded elk."[32]

Health authorities now arranged lectures, published pamphlet-guides, encouraged demonstrations, and carried the latest medical information to the public schools. Not until 1907, however, did the state legislature appropriate even $2,000 for the state board's use in its antituberculosis propaganda campaign. Teacher training in the basic facts of tuberculosis had a double purpose, for a large number of women in the

[32] Dr. Theron A. Wales as quoted in "Quacks in Los Angeles," *Southern California Practitioner*, II (1887), 145.

early stages of phthisis had come to California, and as fairly little else was open to them, they had become primary teachers. A medical man observed that the danger of infection from child to child was less than that from a tuberculous teacher. He recommended regular physical examinations—and "spittoons" in every classroom! Other doctors called for disinfection of school buildings from time to time.

In 1907 the California Association for the Study and Prevention of Tuberculosis was organized and merged with the Southern California Anti-Tuberculosis League. This latter society had urged the establishment the year before of the Los Angeles Helping Station for Indigent Consumptives, an advisory and clinical organization. From 1903 to 1906 the league had distributed thousands of pamphlets to southern California homes and furnished lecturers to various civic groups. Education was effective. As early as 1896 a physician traveling on the streetcar between Los Angeles and Pasadena had noticed that "consumptive looking passengers take from their pockets neat little receptacles in which they expectorated."[33]

In the new century's opening decade, tuberculosis education and research had been accepted as indispensable to the public welfare. Christmas seals first introduced in Denmark in 1904 were adopted in this country three years later. In 1910 Dr. George H. Kress of Los Angeles remarked, "Up to the present time, California has nothing to be proud of as regards aid in the solution of the tuberculosis problem."[34] He could not have repeated his statement with accuracy even half a decade later, for he was speaking of an era virtually past. True, California was not so far advanced in providing the state-sponsored institutions and aid for the tuberculous which New York, Pennsylvania, and Michigan had inaugurated, yet the political vacuum for the concern of invalids had ended tardily but completely.

Unfortunately, health seeking as a significant factor in southern California had almost vanished, too. That disorganized, unplanned move-

[33]"The Fate of a Famous Health Resort," *Southern California Practitioner*, XI (1896), 69.

[34]California Association for the Study and Prevention of Tuberculosis, *Bulletin*, III (1910), 5.

ment probably would not have grown so rapidly under state regulation, had such been possible or desirable. Neither would have such dramatic and often haphazard movements as the gold rush and numerous land booms. Individualistic, ignorant, and idealistic as the pioneering health-seeking migration was, it gained a significance far greater than government planning could have given it.

CHAPTER XI

Climax and Anticlimax

DURING THE last years of the nineteenth century there were probably numerous southern Californians who believed that perpetual motion of a sort had at last been discovered in the migration of health seekers, with hope the inexhaustible fuel and a benign climate the everlasting mainspring of this unusual movement. Yet, if anyone thought the phenomenon was to be permanent, he forgot that it was an entirely human institution and therefore inevitably ephemeral.

By several factors the westward movement for health was slowed and minimized. Many who otherwise would have sought California's climate for a cure were discouraged by the tuberculosis scare of the nineties and following years. Yet this psychological reaction was only the culmination of a growing conviction within California that the health seeker as such had about worn out his welcome. By the 1890's there was graphic evidence of this new attitude. No longer did the press cater to the sick in news stories or in advertisements. Isolated exceptions, of course, did exist, but most writers felt that southern California's sickly element was an annoyance, to be advised how to recover health and not to be a danger to others.[1] "Don'ts" now prevailed over "dos." Economically, the

[1] In 1907 David Starr Jordan noted that the health rush had greatly diminished. He observed: "I know of few things more pitiful than the annual migration of hopeless consumptives which formerly took place to Los Angeles, Pasadena, and San Diego. The Pullman cars in the winter used to be full of sick people, banished from the East by physicians who do not know what else to do with their incurable patients." *California and the Californians*, p. 43.

region could well afford to discourage any unlimited migration of health seekers, since the influx of other groups such as tourists, investors, and workingmen had become so large and the local economy had so broadened that southern California no longer needed to depend on wealthy semi-invalids for a financial advantage. Of course, their contributions were always welcome, but tubercular immigrants as a whole were discouraged without great concern about the possibility of losing those of "the better class."[2]

In the meantime, Arizona was experiencing a new boom in the role of a health resort. To the untraveled easterner of the seventies and eighties, whose knowledge of the Far West was largely limited to newspaper accounts, the sparsely populated country had seemed socially unhealthy, even for a robust white man. Wild Apaches symbolized the territory of those days more than did sanitariums. With better transportation and the pacification of the Indians in the eighties, Arizona was opened as a health resort. Such towns as Sunnyslope, north of Phoenix, were founded by tuberculars. Colorado, which had first become a health resort in the early 1870's, particularly in the Colorado Springs area, enjoyed an awakening when larger numbers of middle westerners began seeking its scenic mountains and pure atmosphere for health. Proximity to large centers of population was also a factor here. Thus southern California had lost its virtual monopoly of the invalid trade.

The greatest factor of all in the decline of the West Coast health resorts was the rapid advances made in medicine. Tuberculars could be about as well, and in some cases even better, treated at sanitariums in their own locality as they could in any alleged climatic Eden. Slowly doctors of the East and Middle West wisely began to discourage their patients from undertaking the traditional western migration. Yet the

[2] In 1905 Dr. C. H. Alden, who had just visited southern California's notable sanitariums, believed that the numbers of tuberculars coming west were not diminishing but were only less noticeable than before because of the larger numbers of "home seekers" then arriving. "The notice formerly posted up, 'Guests will please take their medicine in their rooms, and not in the dining room or parlor,' is no longer needed, as the best resorts will not admit consumptives, and so advertise," he said. "Some Southern California Health Resorts, 1904-5," *Transactions* of the American Climatological Association, XXI (1905), 45-46.

health legend died hard. Gradually the tuberculars arrived in smaller and smaller numbers, leaving a void which no other group of health seekers could ever quite fill. Before 1910 the proportion of invalids among the tourists had already very perceptibly declined both in real and relative numbers at the same time that the general immigration to California began to increase as never before with the opening of the automobile age. California's story from Juan Rodríguez Cabrillo to the space age can be told by tracing trends in migrations. Now, the health rush had completed its significant part in this drama of destiny.

No census was ever made of health seekers to determine how many actually arrived in California. At the beginning of the movement there were no statisticians to keep records, and at the close of the era, about 1900, few tuberculars wanted to advertise their condition. There is no precise knowledge of what ratio southern California's sickly migrants bore to the state's total population. The estimate of 75 per cent made by one doctor is surely a gigantic exaggeration made by an individual virtually surrounded during his every active moment by these people. Roughly a quarter of all southern California's inhabitants was an approximation favored by several newspapers, but Dr. Sumner J. Quint, who served as Los Angeles police surgeon and made up mortality rates and contagion charts, believed that in the key city of the health-resort area tubercular transients numbered about 8 to 10 per cent of the community's population. Dr. John C. King's exact statistics for Banning, which show that 12 per cent of the town's residents were tuberculous, seem to verify such a calculation. Of course, the rheumatics, asthmatics, and sufferers of kidney diseases were numerous but never enumerated. Time and rumor may have overemphasized the role of tuberculars. No one seems ever to have made any guess as to just how many former health seekers there were in southern California during these significant years. Certainly statistics of the pleasanter sort do reveal many cures from tuberculosis when the patients were wise enough to come in the early stage, but usually in a majority of cases when the victims arrived late fatalities occurred. More than all the counsels of the period's doctors these statistics prove that the climate was no medicine. If the sunshine and equable temperatures were not in themselves cure-alls, at least the climate was not negative in its effect, as that of other areas sometimes proved to be. Medically, then, the health-resort movement was not a great tragedy but rather a qualified success story, not too bad a record after all in a world where total victory is rare.

Historically, the health seekers were of much more consequence than they ever were to be in the annals of local medicine. These people ranged economically from the very poorest and most obscure of men to the most powerful millionaires and a few world-famous figures. Geographically they represented every continent, with most of the foreign element from Great Britain, western Europe, and Canada. Still, the great majority were Americans, and although every state and territory, and especially their larger towns, had transplanted citizens on the Pacific Coast, the average southern California health seeker was typically American—middle-class, middle-aged, and middle western. Some localities, however, consciously or unconsciously "specialized" in certain groups of migrants. As we have already learned, Santa Barbara was predominantly New England in taste and origin, Pasadena and Los Angeles had their Hoosiers and "Windy City" migrants, and former Iowans were numerous at Long Beach. Although in the early days after the gold rush San Diego had a decidedly "Southern complexion," this trend was later changed, to some extent by succeeding waves of health seekers.

The phenomenon which these courageous people, whatever their age, origin, or condition, had created significantly changed southern California, helping to provide the semi-isolated region with its first predominance in Anglo-Saxon Americans, a solid basis of family life and social stability, and offering the outpost area its initial experience in large-scale advertising. True, agricultural possibilities were propagandized at the same time, but the appeal of the agrarian campaign was also an integral part of climate and health-resort boosting.[3] In some cases, health seekers founded new towns, such as Pasadena, Sierra Madre, Altadena,

[3]At the height of the rush, even department stores advertised in a vein calculated to win the health seeker. B. F. Coulter of Los Angeles proclaimed: "*Invalids* and all others who wish to protect their systems, should buy the Los Angeles Woolen Mills' Flannels and Underwear." Los Angeles *Express*, Aug. 29, 1883.

Palm Springs, Riverside, the Ojai Valley communities, Palms, and to a large degree even Santa Monica. They named others: for instance, Mentone, Nordhoff, and Carlsbad. One hardly thinks of these places as memorials, but most effectively they were, if now their origins are only half-forgotten souvenirs of a time when southern California was the West's highly publicized "Sanitarium Belt."

By 1900 the search for health in California had reached a climax. There was to be an anticlimax without the usual elements of disappointment. The migration of health seekers did not die out with the new century, nor could it even remain dormant. Apparently, it is unending. The movement has lived on as a minor factor of western life, as have gold mining and fur trapping. Like these once important and romantic undertakings in developing the West, its methods have changed, and changed significantly. The tubercular no longer looks to southern California as his special sanctuary, but even larger numbers of the aged and sick than ever came earlier still come west to lengthen their lives and lessen aches. Health resorts have developed in numbers, varieties, and methods not known before 1900. So have many faith and mind cures. Most important of all, southern California has become one of the nation's and the world's great medical centers. Los Angeles alone possesses a county hospital with over 120 structures covering 56 acres. Merely a list of great hospitals reflects progress. Medical schools and their installations, too, distinguish the region. One particularly thinks of U.S.C.'s school of medicine and U.C.L.A.'s medical center. Tuberculosis still receives considerable attention locally, and the state contains nearly thirty tuberculosis hospitals. Of course, the health seeker of the nineteenth century did not bring about all this more recent medical advancement, but both he and it are a part of the region's history because of one prime resource—climate. The cheerful old story about cool summer nights and warm winter days never tarnishes and continues to lure. That God-given blessing of climate, like the hope it stimulates in the human breast, is virtually eternal.

California, from the beginning a land of hope, has often given those with faith even more than they have expected. In the case of the pioneer health seekers, willing to adapt to their new western home, it meant more than the mere existence they sought. Southern California offered them a freer life and a broader livelihood and the opportunity to build up a promising region as part of the American future.

Bibliography

MANUSCRIPTS

D. M. Berry papers, 1873-1875. The Huntington Library.

Jeanne C. Carr manuscript articles on Pasadena, 1876-1877. The Huntington Library.

Elizabeth Chase. "A History of Ojai Valley." Unpublished Master of Arts thesis, University of Southern California, 1933.

Herbert Crouch diaries, 1868-1917. The Huntington Library.

Charles Dwight Willard letters, 1879-1913. The Huntington Library.

BOOKS, PAMPHLETS, AND ARTICLES

Abbott, Wyllys S. "Los Angeles and the Teachers," *Overland Monthly*, 2nd Ser., XXXIV (1899), 74-83.

Adams, Emma H. *To and Fro in Southern California, with Sketches in Arizona and New Mexico*. Cincinnati, Ohio, 1887.

"Agriculture as an Occupation for Women in California," *Overland Monthly*, 2nd Ser., IX (1887), 652-658.

Aiken, Charles Sedgwick. *California Today, San Francisco, Its Metropolis*. San Francisco, 1903.

Alden, C. H. "Some Southern California Health Resorts, 1904-5," *Transactions* of the American Climatological Association, XXI (1905), 44-54.

All about Santa Barbara, Cal.: The Sanitarium of the Pacific Coast. Santa Barbara, Calif.: Daily Advertizer Printing House, 1878.

Anderson, Winslow. *Mineral Springs and Health Resorts of California with a Complete Chemical Analysis of Every Important Mineral Water in the World*. San Francisco, 1890.

Atchison and Eshelman. *Los Angeles Then and Now*. Los Angeles, 1897.

Ayers, James J. *Gold and Sunshine: Reminiscences of Early California*. Boston, 1922.

Ball, Charles Dexter. *Orange County Medical History*. Santa Ana, Calif., 1926.

Barlow, W. Jarvis. "Report on Two Hundred Charity Cases of Pulmonary Tuberculosis, under Sanatorium Treatment at Los Angeles," *Transactions* of the American Climatological Association, XXIII (1907), 162-167.

Barrett, James Wyman. *Joseph Pulitzer and His World*. New York, 1941.

Bartlett, Dana Webster. *The Better City: A Sociological Study of a Modern City*. Los Angeles, 1907.

Bell, William Abraham. *New Tracks in North America: A Journal of Travel and Adventure whilst Engaged in the Survey for a Southern Railroad to the Pacific Ocean during 1867-8*. 2 vols. London, 1869.

Bentley, William R. *Bentley's Hand-Book of the Pacific Coast*. Oakland, Calif., 1884.

Bicknell, Edmund, comp. *Ralph's Scrap Book*. Lawrence, Mass., 1905.

Bidwell, John. *A Journey to California, with Observations about the Country, Climate and the Route to This Country*. San Francisco, 1937.

"Big Rock Villa," *Sunset*, IV (1899), 19.

Bingham, Edwin R. *Charles F. Lummis, Editor of the Southwest*. San Marino, Calif., 1955.

Bishop, William Henry. *Old Mexico and Her Lost Provinces: A Journey in Mexico, Southern California, and Arizona, by Way of Cuba*. New York, 1883.

Blair, Thomas Arthur. *Climatology, General and Regional*. New York, 1942.

Blodget, Lorin. *Climatology of the United States and of the Temperate Latitudes of the North American Continent*. Philadelphia, 1857.

Bowen, Edith Blumer, comp. *Annals of Early Sierra Madre*. Sierra Madre, Calif., 1950.

Brace, Charles Loring. *The New West; or, California in 1867-1868*. New York, 1869.

Braly, John Hyde. *Memory Pictures: An Autobiography*. Los Angeles, 1912.

Brandt, Lilian, comp. *A Directory of Institutions and Societies Dealing with Tuberculosis in the United States and Canada*. New York, 1904.

Bridge, Norman. *Fragments and Addresses*. Los Angeles, 1915.

———. "How Far Shall the State Restrict Individual Action of the Sick, Especially the Tuberculous?" *California and Western Medicine*, I (1903), 179-182.

———. *The Marching Years*. New York, 1920.

———. *Tuberculosis*. Philadelphia, 1903.

Briggs, Lloyd Vernon. *California and the West, 1881 and Later*. Boston, 1931.

Brook, Harry Ellington. *The Land of Sunshine: Southern California*. Los Angeles, 1893.

Brown, John, Jr., and James Boyd, eds. *History of San Bernardino and Riverside Counties*. [Madison, Wis.], 1922.

Browne, John Ross. *Adventures in the Apache Country: A Tour through Arizona and Sonora, with Notes on the Silver Regions of Nevada*. New York, 1869.

Bryant, Edwin. *What I Saw in California: Being the Journal of a Tour, by the Emigrant Route and South Pass of the Rocky Mountains, across the Continent of North America, the Great Desert Basin, and through California, in the Years 1846, 1847.* New York, 1848.

Bullard, Frank D. "Climatology and Diseases of Southern California," *Southern California Practitioner*, V (1890), 201-220.

Burton, George Ward. *Beloved California: A Lyric of the Soul.* Los Angeles, 1914.

———. *Burton's Book on California and Its Sunlit Skies of Glory.* 3 vols. in 1. Los Angeles, 1909.

[———]. *Men of Achievement in the Great Southwest:... A Story of Pioneer Struggles during Early Days in Los Angeles and Southern California.* Los Angeles, 1904.

Buxton, Edward. "How Invalids Should Come to Southern California," *Southern California Practitioner*, IX (1894), 327-332.

California as It Is. 3rd ed. San Francisco: San Francisco Call Company, 1882.

"The California Out-of-Doors Health Home and School," *Out West*, N.S., IV (1912), 101-111.

Carr, Jeanne C. "Pasadena, and the Story of Rancho San Pascual," in *An Illustrated History of Los Angeles County, California.* Chicago, 1889. Pp. 313-318.

Carroll, H. K. *Report on Statistics of Churches in the United States... 1890.* Washington, D.C., 1894, in U.S. Census Office, 11th Census, 1890. *Census Reports.* Vol. 9.

Caughey, John Walton. *California.* New York, 1940.

———. "Don Benito Wilson: An Average Southern Californian," *Huntington Library Quarterly*, II (1939), 285-300.

Chadwick, Henry Dexter, and Alton Stackpole Pope. *The Modern Attack on Tuberculosis.* New York, 1942.

Chamberlain, W. M. "Notes on the Climatic and Sanitary Conditions of Southern California," *Southern California Practitioner*, I (1886), 485-497.

Chapin, Lon F. *Thirty Years in Pasadena, with an Historical Sketch of Previous Eras.* Vol. I. [Los Angeles], 1929.

Chittenden, Newton H. *Health Seekers', Tourists' and Sportsmen's Guide to the Sea-Side, Lake-Side, Foothill, Mountain and Mineral Spring Health and Pleasure Resorts of the Pacific Coast.* 2nd ed. San Francisco, 1884.

Church, W. B. "Consumption," *California Medical Journal*, XX (1899), 308-309.

Clark, Susie C. *The Round Trip from the Hub to the Golden Gate.* Boston, 1890.

[Coan, T. M.]. "Climates for Invalids," *Harper's Monthly Magazine*, LVIII (1879), 583-589.

Codman, John. *The Round Trip by Way of Panama through California, Oregon, Nevada, Utah, Idaho, and Colorado.* New York, 1879.

Cole, George Llewllyn. *Medical Associates of My Early Days in Los Angeles.* Los Angeles, 1930.

Cook, A. J. "California: Foul Brood Law," *American Bee Journal*, XLV (1905), 325-326.

Cowan, John L. "Climate and Consumption," *Overland Monthly*, 2nd Ser., LV (1910), [244]-[253].

Crafts, Eliza P. R. *Pioneer Days in the San Bernardino Valley*. Redlands, Calif., 1906.

[Croswell, M. S.]. "At San Diego and the Gold-Mines," *Overland Monthly*, V (1870), 320-326.

Dakin, Edwin Franden. *Mrs. Eddy: The Biography of a Virginal Mind*. New York, 1930.

Dall, Caroline H. *My First Holiday; or, Letters Home from Colorado, Utah, and California*. Boston, 1881.

Delamere, H. S. "The Over-Production of Doctors," *California and Western Medicine*, VI (1908), 269-271.

Delano, Alonzo. *Life on the Plains and among the Diggings*. Auburn, N.Y., 1854.

Dennis, Charles Henry. *Eugene Field's Creative Years*. Garden City, N.Y., 1924.

Description of the Advantages Which Pomona Offers Tourists, Invalids and Speculators. Oakland, Calif., 1887.

[Dodge, Mary Abigail]. *Biography of James G. Blaine*. Norwich, Conn., 1895.

Dukeman, William H. "On the Subjection of Consumption," *Pacific Medical Journal*, XXXIV (1891), 22-27.

Dumke, Glenn S. *The Boom of the Eighties in Southern California*. San Marino, Calif., 1944.

Eames, Ninetta. "Bee Culture in California," *Overland Monthly*, 2nd Ser., XVII (1891), [113]-130.

Eddy, Mary Baker. *Science and Health, with Key to the Scriptures*. Boston, 1934.

Edwards, William A., and Beatrice Harraden. *Two Health-Seekers in Southern California*. Philadelphia, 1897.

Elliott, J. H. "The Mortality from Tuberculosis in the Neighborhood of Sanatoriums," *Transactions* of the American Climatological Association, XXII (1906), 280-286.

Ellis, H. Bert. "Blackening Our Own Character," *Southern California Practitioner*, IV (1889), 18-21.

———. "Population of Los Angeles," *Southern California Practitioner*, IX (1894), 233-234.

Esperanza: The Sierra Home of the Open Air Cure. Altadena, Calif., 1904.

Farnsworth, R. W. C., ed. *A Southern California Paradise, (in the Suburbs of Los Angeles.): Being a Historic and Descriptive Account of Pasadena, San Gabriel, Sierra Madre, and La Cañada*. Pasadena, Calif., 1883.

"The Fate of a Famous Health Resort," *Southern California Practitioner*, XI (1896), 69-70.

Ferrell, John Atkinson, and Pauline A. Mead, comps. *History of County Health Organizations in the United States, 1908-33*. Washington, D.C., 1936.

Finck, Henry T. *The Pacific Coast Scenic Tour, from Southern California to Alaska, the Canadian Pacific Railway, Yellowstone Park and the Grand Cañon*. New York, 1890.

Folsom, E. C. "Gaseous Enemata in Consumption," *Southern California Practitioner*, II (1887), 235-236.

Frank, Herman W. *Scrapbook of a Western Pioneer*. Los Angeles, 1934.

Frémont, Jessie Benton. *Far-West Sketches*. Boston, 1890.

Gibbons, Henry, Sr. "The Inheritance of Pulmonary Disease: Its Possible Eradication, with Especial Reference to the Climate of San Diego," *Pacific Medical Journal*, XXIII (1881), 403-406.

———. "Where Shall We Send Our Consumptive Patients?" *Pacific Medical Journal*, N.S., IV (1870), 312-313.

Giddings, Jennie Hollingsworth. *I Can Remember Early Pasadena*. Los Angeles, 1949.

Gilman, Charlotte Perkins. *In This Our World*. Boston, 1898.

———. *The Living of Charlotte Perkins Gilman: An Autobiography*. New York, 1935.

Graham, Margaret Collier. *Do They Really Respect Us? and Other Essays*. San Francisco, 1912.

Gresham, Matilda. *Life of Walter Quintin Gresham, 1832-1895*. 2 vols. Chicago, 1919.

Guinn, James M. *Historical and Biographical Record of Southern California*. Chicago, 1902.

———. *A History of California and an Extended History of Its Southern Coast Counties*. 2 vols. Los Angeles, 1907.

Gunn, Douglas. *Picturesque San Diego, with Historical and Descriptive Notes*. Chicago, 1887.

Hagadorn, Wesley. "A Pasadena Letter," *Southern California Christian Advocate* (Los Angeles), Aug. 15, 1887, p. 5.

Hall-Wood, Mary C. F. *Santa Barbara as It Is: Topography, Climate, Resources, and Objects of Interest*. Santa Barbara, Calif., [1884].

Hambaugh, J. M. "California the Land for the Sick," *Gleanings in Bee Culture*, XXVI (1898), 48-49.

Harraden, Beatrice. "Some Impressions of Southern California," *Blackwood's Edinburgh Magazine*, CLXI (1897), 172-180.

Harris, Henry. *California's Medical Story*. San Francisco, 1932.

Harrison, Carter Henry, II. *Stormy Years: The Autobiography of Carter H. Harrison, Five Times Mayor of Chicago*. Indianapolis, Ind., 1935.

Hastings, Lansford W. *The Emigrants' Guide to Oregon and California*. Cincinnati, Ohio, 1845; reprinted, Princeton, N. J., 1932.

Hicks, Ratcliffe. *Southern California, or The Land of the Afternoon*. Springfield, Mass., 1898.

Hine, Robert V. *California's Utopian Colonies*. San Marino, Calif., 1953.

History of San Bernardino County, California. San Francisco, 1883.

History of San Diego County, California. San Francisco, 1883.

Holder, Charles Frederick. *All about Pasadena and Its Vicinity*. Boston, 1889.
Holton, Edward D. *Travels with Jottings: From Midland to the Pacific*. Milwaukee, Wis., 1880.
Hornbeck, Robert. *Roubidoux's Ranch in the 70's*. Riverside, Calif., 1913.
Hunt, Nancy A. "By Ox-Team to California," *Overland Monthly*, 2nd Ser., LXVII (1916), [317]-326.
Hutchinson, Woods. *The Conquest of Consumption*. Boston, 1910.
An Illustrated History of Southern California. Chicago, 1890.
Ingraham, Charles Wilson. "Control of Tuberculosis from a Strictly Medico-Legal Standpoint," *Journal*, American Medical Association, XXVII (1896), 693.
Inside Facts concerning the Pasadena and Mt. Wilson Railway Company. Pasadena, 1897.
Ives, Sarah Noble. *Altadena*. Pasadena, Calif., 1938.
Jackson, Abraham Willard. *Barbariana: or Scenery, Climate, Soils and Social Conditions of Santa Barbara City and County, California*. San Francisco, 1888.
Jackson, Helen Hunt. *Glimpses of California and the Missions*. Boston, 1902.
James, George Wharton. *B. R. Baumgardt & Co's Tourists' Guide Book to South California*. Los Angeles, 1895.
———. *Heroes of California: The Story of the Founders of the Golden State as Narrated by Themselves or Gleaned from Other Sources*. Boston, 1910.
———."How To Live Out-of-Doors," *Life and Health*, XXV (1910), 470-477.
———. *The Influence of the Climate of California upon Its Literature*. Los Angeles, [190_?].
———. *Rose Hartwick Thorpe and the Story of "Curfew Must Not Ring To-Night."* Pasadena, Calif., 1916.
———. *The Wonders of the Colorado Desert (Southern California): Its Rivers and Its Mountains, Its Canyons and Its Springs, Its Life and Its History*. 2 vols. Boston, 1906.
Johnston, Charles William. *Along the Pacific by Land and Sea, through the Golden Gate*. Chicago, 1916.
Johnston, Nathan Robinson. *Looking Back from the Sunset Land; or, People Worth Knowing*. Oakland, Calif., 1898.
Jordan, David Starr. *California and the Californians*. San Francisco, 1907.
Kearney, Arthur. *San Bernardino County: Its Resources and Climate*. San Bernardino, Calif., 1874.
Kenderdine, Thaddeus S. *California Revisited, 1858-1897*. Newtown, Pa., 1898.
King, John C. "A Review of What Has Been Done for the Prevention of the Spread of Tuberculosis in the State of California," *California and Western Medicine*, IV (1906), 164-166.
Kinsley, Philip. *The Chicago Tribune: Its First Hundred Years*. Vols. I-III. Chicago, 1943-1946.
Knopf, S. A. "The California Quarantine against Consumptives," *Forum*, XXVIII (1900), 615-620.

Kress, George Henry. *A History of the Medical Profession of Southern California.* Los Angeles, 1910.

———. "Is the Health of Los Angeles Menaced by Pulmonary Tuberculosis?" *California and Western Medicine,* III (1905), 207-209.

———. "Tuberculosis," *Los Angeles Medical Journal,* I (1904), 314-321.

Kuhrts, Jacob. "Reminiscences of a Pioneer," *Publications,* Historical Society of Southern California, VII (1906), [59]-68.

Lamb, Martha J. "Chief Justice Morrison Remick Waite," *Magazine of American History,* XX (1888), 1-16.

Leech, Harper, and John Charles Carroll. *Armour and His Times.* New York, 1938.

Lester, John Erastus. *The Atlantic to the Pacific.* London, 1873.

Leuba, Edmond. *La Californie et les états du Pacifique: Souvenirs et impressions.* Paris, 1882.

Lewis, Dio. *Gypsies; or, Why We Went Gypsying in the Sierras.* Boston, 1881.

Lindley, Walter, and Joseph P. Widney. *California of the South.* 3rd ed. New York, 1896.

[Lindley, Walter]. "Riverside and San Bernardino: An Editorial Trip," *Southern California Practitioner,* I (1886), 376-378.

———. *Ships That Never Reached the Harbor.* Los Angeles, 1921.

Lippincott, Sara Jane Clarke. *New Life in New Lands: Notes of Travel.* New York, 1873.

Los Angeles Board of Supervisors. "Petition of the Board of Supervisors for an Appropriation for the Support of the Non-Resident Indigent Sick of Los Angeles County," in *Appendix to Journals of Senate and Assembly, of the Eighteenth Session of the Legislature of the State of California.* Vol. III. Sacramento, 1870.

The Los Angeles Case: The People vs. Merrill Reed, et al. Boston, [1909].

Los Angeles Chamber of Commerce. *Facts and Figures concerning Southern California and Los Angeles City and County.* Los Angeles, 1888.

Los Angeles County Pioneer Society. *Historical Record and Souvenir.* Los Angeles, 1923.

Lows, O. S. "Another View of the [Tuberculosis] Question," *California Medical Journal,* XX (1899), 309-310.

[Ludwig Salvator, Archduke of Austria]. *Eine Blume aus dem goldenen Lande oder Los Angeles.* Prag, 1878.

———. *Los Angeles in the Sunny Seventies,* trans. Marguerite Eyer Wilbur. Los Angeles, 1929. (Translation of *Eine Blume.*)

Lummis, Charles F. *A Tramp across the Continent.* New York, 1892.

McCarthy, John Russell. "Joseph P. Widney's 'Century of Service,'" *California History Nugget,* VII (1939), 35-39.

McGrew, Clarence Alan. *City of San Diego and San Diego County, the Birthplace of California.* 2 vols. Chicago, 1922.

McGroarty, John Steven, ed. *History of Los Angeles County.* 3 vols. Chicago, 1923.

———. *Los Angeles from the Mountains to the Sea.* 3 vols. Chicago, 1921.

McIntyre, J. F. "California as a Bee-Keeping State," *Gleanings in Bee Culture,* XXIII (1895), 862.

McPherson, William. *Homes in Los Angeles City and County and Description Thereof.* Los Angeles, 1873.

McWilliams, Carey. *Southern California Country: An Island on the Land.* New York, 1946.

Magee, Harvey White. *The Story of My Life.* Albany, N.Y., 1926.

[Mason, Jesse D.]. *History of Santa Barbara County, California.* Oakland, Calif., 1883.

Mast, Isaac. *The Gun, Rod and Saddle; or, Nine Months in California.* Philadelphia, 1875.

"The Mecca of the Quack," *Los Angeles Medical Journal,* II (1905), 19-20.

Miller, Lizzie E. "Letter from California," *Southern California Christian Advocate* (Los Angeles), Jan. 30, 1887, p. 14.

Modjeska, Helena. *Memories and Impressions of Helena Modjeska: An Autobiography.* New York, 1910.

Muir, John. "The Bee Pastures of California," *Century Illustrated Monthly Magazine,* XXIV (1882), 222-229, 388-396.

Munk, Joseph Amasa. *Descriptive Climatology of the Southwest.* Indianapolis, Ind., 1908.

———. "Impressions of Southern California," *California Medical Journal,* XVI (1895), 174-176.

———. "Some Defects of Our Climate," *California Medical Journal,* XIX (1898), 148-150.

Newmark, Harris. *Sixty Years in Southern California, 1853-1913.* New York, 1926.

[Nichols, C. S.]. *The Prolific Seven: Where To Find Health, Wealth and Pleasure.* Los Angeles, 1895.

Nordhoff, Charles. *California: For Health, Pleasure, and Residence.* New York, 1873.

Norton, A. "California for Consumptives: Not the Most Favorable Region in the World for That Disease; Some Interesting Facts for Health-Seekers," *Gleanings in Bee Culture,* XXVI (1898), 303-304.

Nottage, Charles George. *In Search of a Climate.* London, 1894.

Pasadena Board of Trade. *Pasadena, Its Climate, Homes, Resources, Etc., Etc.* Buffalo, N.Y., 1888.

Paulson, Luther L., comp. *Hand-Book and Directory of San Luis Obispo, Santa Barbara, Ventura, Kern, San Bernardino, Los Angeles & San Diego Counties.* San Francisco, 1875.

Pellett, Frank Chapman. *History of American Beekeeping.* Ames, Iowa, 1938.

Perkins, Joseph J. *A Business Man's Estimate of Santa Barbara County, California.* Santa Barbara, Calif., 1881.

Perling, Joseph Jerry. *Presidents' Sons: The Prestige of Name in a Democracy.* New York, 1947.

Phillips, David L. *Letters from California: Its Mountains, Valleys, Plains, Lakes, Rivers, Climate and Productions.* Springfield, Ill., 1877.

Phillips, Michael James. *History of Santa Barbara County, California, from Its Earliest Settlement to the Present Time.* 2 vols. Chicago, 1927.

Pottenger, Francis Marion. *The Fight against Tuberculosis: An Autobiography.* New York, 1952.

———. "Is Another Chapter in Public Phthisiphobia about To Be Written?" *California and Western Medicine,* I (1903), 81-84.

Power, Bertha Knight. *William Henry Knight: California Pioneer.* [New York], 1932.

Powers, Stephen. *Afoot and Alone: A Walk from Sea to Sea by the Southern Route.* Hartford, Conn., 1872.

Praslow, J. *The State of California: A Medico-Geographical Account (Goettingen, 1857),* trans. Frederick C. Cordes. San Francisco, 1939.

Pringle, Henry F. *The Life and Times of William Howard Taft.* 2 vols. New York, 1939.

"Quacks in Los Angeles," *Southern California Practitioner,* II (1887), 144-145.

Reid, Hiram Alvin. *History of Pasadena, Comprising an Account of the Native Indian, the Early Spanish, the Mexican, the American, the Colony and the Incorporated City.* Pasadena, Calif., 1895.

Remondino, Peter C. *The Mediterranean Shores of America: Southern California: Its Climatic, Physical, and Meteorological Conditions.* Philadelphia, 1892.

———. "Physical Conditions and Meteorology of Southern California," *Southern California Practitioner,* VI (1891), 1-45.

Rice, George. *Southern California Illustrated.* Los Angeles, 1883.

Roberts, Edwards. *Santa Barbara and around There.* Boston, 1886.

Rochester, DeLancey. "The Role of Local Sanatoria in Preventing the Spread of Tuberculosis," *Transactions* of the American Climatological Association, XIX (1903), 60-63.

Roorbach, Eloise. "Making Pottery on the California Hills," *Craftsman,* XXIV (1913), 343-346.

Root, E. R. "Notes on Travel," *Gleanings in Bee Culture,* XXX (1902), 288-291.

Rusling, James F. *The Great West and Pacific Coast.* New York, 1877.

Russell, Richard Joel. *Climates of California.* (University of California Publications in Geography, II, No. 4). Berkeley, Calif., 1926.

Sanborn, Katherine Abbott. *A Truthful Woman in Southern California.* New York, 1893.

Sanders, Francis Charles Scott. *California as a Health Resort.* San Francisco, 1916.

San Diego Chamber of Commerce. *Descriptive, Historical, Commercial, Agricultural, and Other Important Information Relative to the City of San Diego, California.* San Diego, Calif., 1874.

Sands, Frank. *Santa Barbara at a Glance*. Santa Barbara, Calif., 1895.
Santa Barbara Board of Trade. *Santa Barbara as a Summer Resort*. Santa Barbara, Calif., 1895.
Saunders, Charles Francis. *The Story of Carmelita: Its Associations and Its Trees*. Pasadena, Calif., 1928.
———. *Under the Sky in California*. New York, 1913.
Seitz, Don Carlos. *Joseph Pulitzer: His Life & Letters*. New York, 1924.
"Sensational Statements regarding Consumption," *Journal*, American Medical Association, XXVII (1896), 107.
Shaw, Pringle. *Ramblings in California*. Toronto, [1857?].
Sherer, John Calvin. *History of Glendale and Vicinity*. Glendale, Calif., 1922.
Sherman, John. *John Sherman's Recollections of Forty Years in the House, Senate and Cabinet*. 2 vols. Chicago, 1895.
Shugart, K. D. "Climatology and Hygiene," *Southern California Practitioner*, VI (1891), 287-296.
Shuman, John W. *California Medicine: A Review*. [Los Angeles], 1930.
Smith, Clifford Pabody. *Historical Sketches, from the Life of Mary Baker Eddy and the History of Christian Science*. Boston, 1941.
Smith, Walter Gifford. *The Story of San Diego*. San Diego, Calif., 1892.
Smythe, William E. *History of San Diego, 1542-1908*. 2 vols. San Diego, Calif., 1908.
Southern California Bureau of Information. *Southern California: An Authentic Description of Its Natural Features, Resources, and Prospects, Containing Reliable Information for the Homeseeker, Tourist, and Invalid*. Los Angeles, 1892.
Southern California: The Italy of America. [n.p., n.d.].
Southern Pacific Company. *California: Its Attractions for the Invalid, Tourist, Capitalist and Homeseeker*. San Francisco, 1892.
Speir, Robert F. *Going South for the Winter, with Hints for Consumptives*. New York, 1870.
Storey, Samuel. *To the Golden Land: Sketches of a Trip to Southern California*. London, 1889.
Storke, Yda Addis. *A Memorial and Biographical History of the Counties of Santa Barbara, San Luis Obispo and Ventura, California*. Chicago, 1891.
Tevis, A. H. *Beyond the Sierras; or, Observations on the Pacific Coast*. Philadelphia, 1877.
Thompson, Warren S. *Growth and Changes in California's Population*. Los Angeles, 1955.
Thornton, Jessy Quinn. *The California Tragedy*. Oakland, Calif., 1945.
Thrasher, Marion. *Long Life in California*. Chicago, 1915.
"Three Health Resorts: Beaumont, Banning and Palm Springs," *Sunset*, IV (1899), 79-81.
Truman, Benjamin Cummings. *Semi-Tropical California: Its Climate, Healthfulness, Productiveness, and Scenery*. San Francisco, 1874.

———. *Tourists' Illustrated Guide to the Celebrated Summer and Winter Resorts of California Adjacent to and upon the Lines of the Central and Southern Pacific Railroads*. San Francisco, 1883.

United States Bureau of the Census. *Tuberculosis in the United States*. (Prepared for the International Congress on Tuberculosis, Sept. 12 to Oct. 12, 1908). Washington, D.C., 1908.

Vachell, Horace Annesley. *Life and Sport on the Pacific Slope*. London, 1900.

[Van Dyke, Theodore S.]. *The Advantages of the Colony of El Cajon, San Diego County, California*. San Diego, Calif., 1883.

———. "Field Sports in San Diego Co., Cal.," *Forest and Stream, Rod and Gun*, XI (1878), 337-338.

———. *Flirtation Camp: or, The Rifle, Rod, and Gun in California*. New York, 1881.

———. *Millionaires of a Day: An Inside History of the Great Southern California "Boom."* New York, 1890.

———. *Southern California: Its Valleys, Hills, and Streams; Its Animals, Birds, and Fishes; Its Gardens, Farms, and Climate*. New York, 1886.

Wagg, Peter A., [pseud.]. *Southern California Exposed*, ed. Homer Fort. Los Angeles, 1915.

Walker, Franklin D. *A Literary History of Southern California*. Berkeley, Calif., 1950.

Walker, W. S. *Between the Tides: Comprising Sketches, Tales and Poems*. Los Gatos, Calif., 1885.

Walters, Frederick Rufenacht. *Sanatoria for Consumptives in Various Parts of the World*. London, 1899.

Waring, George E., Jr., comp. *Report on the Social Statistics of Cities*. Part II. Washington, D.C., 1887, in U.S. Census Office, 10th Census, 1880. *Census Reports*. Vol. 19.

Warner, Charles Dudley. *On Horseback: A Tour in Virginia, North Carolina, and Tennessee with Notes of Travel in Mexico and California*. Boston, 1889.

———. *Our Italy*. New York, 1891.

———. "The Outlook in Southern California," *Harper's Monthly Magazine*, LXXXII (1891), [167]-189.

Webb, Gerald Bertram. *Tuberculosis*. New York, 1936.

Weeks, George F. *California Copy*. Washington, D.C., 1928.

Whiting, Lilian. *The Land of Enchantment, from Pike's Peak to the Pacific*. Boston, 1906.

Widney, Joseph P. *The Greater City of Los Angeles: A Plan for the Development of Los Angeles City as a Great World Health Center*. Los Angeles, 1938.

Widney, Robert M. *Ontario: Its History, Description, and Resources*. Riverside, Calif., 1884.

Willard, Charles Dwight. *The Herald's History of Los Angeles City*. Los Angeles, 1901.

Wills, Mary H. *A Winter in California*. Norristown, Pa., 1889.
Wilson, Carol Green. "A Business Pioneer in Southern California," *Quarterly*, Historical Society of Southern California, XXVI (1944), 139-161.
Winther, Oscar Osburn. "The Use of Climate as a Means of Promoting Migration to Southern California," *Mississippi Valley Historical Review*, XXXIII (1946), 411-424.
Wood, E. N. *Guide to Santa Barbara, Town and County*. Santa Barbara, Calif., 1872.
Worthington, Henry. "Temecula Hot Springs," *Southern California Practitioner*, II (1887), 205-207.
Zierer, Clifford M. "Migratory Beekeepers of Southern California," *Geographical Review*, XXII (1932), 266-269.

PERIODICALS AND SERIAL PUBLICATIONS

American Bee Journal (Cincinnati, Ohio), XXX-L (1890-1910).
American Climatological Association, *Transactions* (New York), XIX-XXVII (1903-1911).
American Medical Association, *Journal* (Chicago), I-LIV (1883-1910).
Argonaut (San Francisco), 1878.
Barlow Sanatorium, *Annual Report* (Los Angeles), I-XIII (1904-1915).
California and Western Medicine (San Francisco), I-X (1903-1912).
California Association for the Study and Prevention of Tuberculosis, *Bulletin*, (Sierra Madre, Calif.), I-III (1908-1911).
California Cultivator and Poultry Keeper (Los Angeles), I-XI (1887-1897).
California Farmer and Journal of Useful Sciences (San Francisco), 1873-1878.
California Medical Association, *Transactions* (San Francisco), N.S., XXI-XXVIII (1891-1898).
California Medical Gazette (San Francisco), I-II (1868-1870).
California Medical Journal (Oakland), X-XXIX (1889-1908).
California State Agricultural Society, *Transactions*, 1858, 1874, in *Appendix to Journals of the Senate and Assembly... of the Legislature of the State of California* (Sacramento), 10th, 21st Sessions.
California State Board of Health, *Biennial Report*, I-XX (1871-1908), all except 1st in *Appendix to the Journals of the Senate and Assembly... of the Legislature of the State of California* (Sacramento), 20th-38th Sessions.
Christian Science Journal (Boston), I-XXII (1883-1904).
Commonwealth Club of California, *Transactions* (San Francisco), II-V (1906-1910).
Gleanings in Bee Culture (Medina, Ohio), XXIII-XXXIX (1895-1910).
International Congress on Tuberculosis, *Transactions... of the Sixth International Congress* (Philadelphia), 1908.

The Lancet (London), 1870-1900.

The Land of Sunshine (Los Angeles), I-XI (1894-1899).

Los Angeles Board of Trade, *Annual Report* (Los Angeles), 1888-1891.

Los Angeles City Board of Health, *Annual Report* (Los Angeles), 1889-1892.

Los Angeles City Directory (Los Angeles), 1901.

Los Angeles County Health Department, *Annual Report* (Los Angeles), 1890-1900.

Los Angeles Journal of Eclectic Medicine (Los Angeles), I-III (1904-1906).

Los Angeles Medical Journal (Los Angeles), I-III (1904-1906).

Los Angeles Polyclinic (Los Angeles), I-III (1895-1896).

Mining and Scientific Press (San Francisco), LVI (1888).

National Conference of State Boards of Health, *Proceedings* (New York), III, IX (1887, 1893).

Occidental Medical News (Sacramento), III-VII (1889-1893).

Pacific Medical Journal (San Francisco), N.S., I-XI and XXIII, XXXIV (1867-1877, 1881, 1891).

Pacific Rural Press (San Francisco), 1874, 1877-1883.

Pasadena City Directory (Los Angeles and Pasadena), 1893/1894, 1900. (Title varies: *The Moore Pasadena City Directory.*)

The Pioneer (San Jose, Calif.), 1894.

Rural Californian (Los Angeles), 1890.

San Francisco Western Lancet (San Francisco), I-XIII (1872-1884).

Southern California Christian Advocate (Los Angeles), 1886-1892.

Southern California Practitioner (Los Angeles), I-XXV (1886-1910).

Southern California Sanitarian and Climatologist (Redlands, Calif.), I (1895).

U. S. Public Health Service, *Public Health Reports* (Washington, D.C.), XII-XIV (1898-1900).

U. S. Public Health Service, *Transactions of the... Annual Conference of State and Territorial Health Officers with the United States Public Health Service* (Washington, D. C.), 1st-4th Conferences (1903-1906).

NEWSPAPERS

Banning *Herald,* 1891-1893.
Beaumont *Sentinel,* 1889-1890.
Chino *Valley Champion,* 1888-1892.
Citrograph (Redlands), 1889-1890.
Daily San Diegan (San Diego), 1891-1892.
Downey *Champion,* 1891-1896.
Elsinore *Press,* 1891-1893.
Fallbrook *Review,* 1891.
Julian *Sentinel,* 1891.
Los Angeles *Commercial,* 1881-1882.
Los Angeles *Commercial Advertiser,* 1886.
Los Angeles *Daily Journal,* 1893-1895.
Los Angeles *Daily News,* 1864-1873.
Los Angeles *Express,* 1875-1883, 1888-1900.
Los Angeles *Herald,* 1874-1887.
Los Angeles *Republican,* 1876.
Los Angeles *Social World,* 1887.
Los Angeles *Star,* 1871-1877.
Los Angeles *Telegram,* 1893.
Los Angeles *Times,* 1884-1886, 1892-1904.
Los Angeles *Tribune,* 1886-1888.
National City *Record,* 1890-1891.
Needles *Eye,* 1891-1894.
Oceanside *Herald,* 1889.
Ontario *Observer,* 1888-1893.
Pasadena *Chronicle,* 1883-1884.
Pasadena *Daily News,* 1898-1899.
Pasadena *Star,* 1889-1901.
Pasadena *Union,* 1883-1889.
People's Advocate (Ventura), 1893-1894.
Pomona *Progress,* 1886-1891.
Pomotropic (Azusa), 1891-1892.
Porcupine (Los Angeles), 1882-1890.
Le Progrés (Los Angeles), 1897.
Redondo Beach *Compass,* 1892.
Riverside *Daily Press,* 1891.
Riverside *Enterprise,* 1891-1893.
San Bernardino *Daily Courier,* 1889.
San Bernardino *Daily Index,* 1888.
San Bernardino *Daily Times,* 1886-1887.
San Diego *Daily Bulletin,* 1869-1872.
San Diego *Progress,* 1897.
San Diego *Sun,* 1887-1893.
San Diego *Union,* 1868-1877, 1887-1891.
San Diego *World,* 1872-1876.
San Francisco *Chronicle,* 1892.
San Pedro *Times,* 1892.
Santa Ana *Herald,* 1881-1882.
Santa Ana *Standard,* 1890.
Santa Barbara *Independent,* 1891-1896.
Santa Barbara *Press,* 1871-1872, 1876-1877, 1881.
Search Light (Los Angeles), 1894.
Sierra Madre *Vista,* 1888-1890.
Southern California Informant (San Diego), 1887.
South Riverside *Bee,* 1888-1890.
Ventura *Signal,* 1873, 1878, 1881, 1884.
Ventura *Vidette,* 1889-1890.
Ventura *Weekly Free Press,* 1893.
Whittier *Register,* 1903.

Index